# Geological Data Management

Geological Society Special Publications
*Series Editor* A. J. FLEET

GEOLOGICAL SOCIETY SPECIAL PUBLICATION NO. 97

# Geological Data Management

EDITED BY

## J. R. A. GILES
British Geological Survey
Nottingham
UK

1995

Published by

The Geological Society

London

# THE GEOLOGICAL SOCIETY

The Society was founded in 1807 as The Geological Society of London and is the oldest geological society in the world. It received its Royal Charter in 1825 for the purpose of 'investigating the mineral structure of the Earth'. The Society is Britain's national society for geology with a membership of around 8000. It has countrywide coverage and approximately 1000 members reside overseas. The Society is responsible for all aspects of the geological sciences including professional matters. The Society has its own publishing house, which produces the Society's international journals, books and maps, and which acts as the European distributor for publications of the American Association of Petroleum Geologists, SEPM and the Geological Society of America.

Fellowship is open to those holding a recognized honours degree in geology or cognate subject and who have at least two years' relevant postgraduate experience, or who have not less than six years' relevant experience in geology or a cognate subject. A Fellow who has not less than five years' relevant postgraduate experience in the practice of geology may apply for validation and, subject to approval, may be able to use the designatory letters C. Geol. (Chartered Geologist).

Further information about the Society is available from the Membership Manager, the Geological Society, Burlington House, Piccadilly, London W1V 0JU, UK. The Society is a Registered Charity, No. 210161.

Published by The Geological Society from:
The Geological Society Publishing House
Unit 7
Brassmill Enterprise Centre
Brassmill Lane
Bath BA1 3JN
UK
(*Orders*: Tel. 01225 445046
Fax 01225 442836)

First published 1995

The publishers make no representation, express or implied, with regard to the accuracy of the information contained in this book and cannot accept any legal responsibility for any error or omission that may be made.

**British Library Cataloguing in Publication Data**
A catalogue record for this book is available from the British Library.

ISBN 1-897799-34-X

Typeset by Aarontype Ltd, Unit 47, Easton Business Centre, Felix Road, Bristol BS5 0HE, UK

Printed in Great Britain by
The Cromwell Press Ltd, Broughton Gifford, Melksham, Wiltshire

Distributors
USA
  AAPG Bookstore
  PO Box 979
  Tulsa
  OK 74101-0979
  USA
  (*Orders*: Tel. (918) 584-2555
    Fax (918) 584-0469)

Australia
  Australian Mineral Foundation
  63 Conyngham Street
  Glenside
  South Australia 5065
  Australia
  (*Orders*: Tel. (08) 379-0444
    Fax (08) 379-4634)

India
  Affiliated East-West Press PVT Ltd
  G-1/16 Ansari Road
  New Delhi 110 002
  India
  (*Orders*: Tel. (11) 327-9113
    Fax (11) 326-0538)

Japan
  Kanda Book Trading Co.
  Tanikawa Building
  3-2 Kanda Surugadai
  Chiyoda-Ku
  Tokyo 101
  Japan
  (*Orders*: Tel. (03) 3255-3497
    Fax (03) 3255-3495)

**Cover photo.** Reproduced by permission of the Director, British Geological Survey. NERC copyright reserved.

# Contents

# The what, why, when, how, where and who of geological data management

JEREMY R. A. GILES

*British Geological Survey, Kingsley Dunham Centre, Keyworth, Nottingham, NG12 5GG, UK*

**Abstract:** The variety, form and volume of data available to geologists have increased significantly over the past few decades. It has become essential to use databases (either analogue or digital) to turn this avalanche of data into usable information. A significant percentage of databases that are created do not repay the cost and effort of their creation and can therefore be considered to be failures. The probability of creating a successful database can be increased by careful consideration of a few simple questions. What is the objective that the database will meet? Why is a digital database being considered? When, if ever, will the cost and effort of creating the database be repaid? How will the database be designed and created? Where are the users? And finally, who will be responsible for maintaining the integrity of the database?

This book reviews the principle and practice of the creation, management and use of geological databases, from the point of view of geological database designers, geological database managers and users. It is the object of this book to increase the proportion of successful geological database projects by encouraging careful thought about the processes that are involved in the design and management of a database.

During the course of their work geologists collect large volumes of data that may be of diverse types. The dataset is commonly so large, and of such variety, that the information that it contains is obscure. Databases can provide a tool that aids geologists to extract information from datasets. At its simplest, a database is a collection of related data. It makes no difference whether it is in a field note book, a filing cabinet or on the disk of a computer. As long as the collection of data is logically coherent with some inherent meaning, it is a database (Elmasri & Navathe 1989). Only random assortments of data are excluded by this definition. If this viewpoint is accepted, geologists have been building and using databases since William Smith opened his first field note book and wrote down an observation.

During the past two decades geologists have built large and complex digital databases to support their activities. Not all of these database projects have been successful, but failure is not a problem unique to geology. Many high profile and well-funded database projects have failed for a variety of complex reasons. Success can be measured in a variety of ways, ranging from the prestige that accrues to an organization that is seen to use innovative technology, through to a simple monetary equation: has the cost and effort of creating and maintaining the database been repaid in terms of new insights into the data and increased productivity and effectiveness?

When Rudyard Kipling wrote the following lines he knew nothing of the problems of geological database design and management. However, his six servants are no less valuable to us.

> *I keep six honest serving-men*
> *(They taught me all I knew);*
> *Their names are What and Why and When*
> *And How and Where and Who.*

These six 'serving-men' should act as a guide to everyone contemplating creating and then managing a geological database.

- What objective will the database meet?
- Why is a digital database being considered?
- When, if ever, will the cost and effort of creating the database be repaid?
- How will the database be designed and created?
- Where are the users?
- Who will be responsible for maintaining the integrity of the database?

If you plan to build and maintain a digital database you must be able to answer each of these questions clearly and unambiguously. A database is a tool and not an end in itself. There is no value in building a database that might be of some use to someone in the future. It must meet clear needs, not amorphous future hopes.

From Giles, J. R. A. (ed.) 1995, *Geological Data Management*,
Geological Society Special Publication No 97, pp. 1–4.

## What objective will the database meet?

The creation of a digital database is commonly seen as an end in itself, a deliverable of a project. This view is unsatisfactory as data cost a lot to acquire, manage, disseminate, store and use. Such effort can only be justified when the information contained within the database is fundamental to the objectives of the organization or project. In all organizations, but especially scientific organizations, information is pivotal to their activities. The efficient management of the processes that turn data into information must be a fundamental priority. Geological organizations do not differ from any other in this rule. Their special problem is the complexity and diversity of the data that must be managed (**Rasmussen**).

The function of a database is to organize a volume of data and to make it accessible, so that it becomes information. There is no point in doing this if there is no use for the information when the database becomes available. A clear purpose for a database must be defined. **Doorgakant**, for example, clearly states the purpose of the Groundwater Level Archive for England and Wales. It meets a clear objective, supporting the preparation of yearbooks, reports and journals to fulfill a statutory requirement, cost effectively. All geological databases need equally clear objectives.

The objectives should be practical and straightforward, clearly encapsulating a real need. If there is no clear definition of the objective there is a real danger that the database will provide no information to anyone, and become a costly burden to the organization.

## Why is a digital database being considered?

The popular image of the database derives from data-processing activities such as large volumes of data being used in a repetitive and predictable manner to produce bank statements or invoices. There is a common assumption that all databases are digital, but banks and utilities used manual databases to process data for statements and invoices long before electronic computers were invented. Their data-processing operations prior to computers consisted of manual systems and paper records. It would be foolish for banks or utilities to return to using manual systems for data-processing, but manual systems may be entirely appropriate to other tasks. Many who have tried using electronic diaries and address books still find the paper equivalents preferable.

If the data are of limited volume or simple structure, a card index, or similar system, is still a valuable and appropriate tool.

It can be argued that there is no need to build a digital system if the database and its applications are simple, well defined, and not expected to change and if there is no requirement to share the data set between many users. The cost of the initial development and continuing overheads of running the system may never be repaid.

The principal reasons for creating a digital database (Date 1986; Elmasri & Navathe 1989) are:

- digital systems provide the ability to share data between many users
- digital format allows input into visualization and modelling software
- digital systems increase speed of retrievals
- complex self-checking can be built into digital systems
- security is good in a properly managed digital system

One of the biggest potential advantages of digital databases is the ability they provide to share data. Sharing can take a number of forms, but most commonly sharing is in the form of a multi-user database. Sophisticated database management software permits many people to use data simultaneously whilst preventing them from interfering with the actions of other users. Such systems work well on single or geographically close sites, but not on a worldwide basis. **Henley** describes a common problem in the geological sciences. An organization such as a mineral exploration company has many, essentially similar, projects working in diverse localities around the world. If uncontrolled, such a situation can lead to drastic data loss or to incompatibilities in data recording standards. **Henley** advocates the use of common software standards and data dictionaries across an organization. This permits projects some freedom to develop databases to meet their own needs, but key elements of data can easily be incorporated into a strategic central database at the end of the project.

Common software standards on an industry-wide scale are promoted by **Chew** and by **Saunders** *et al*. The big advantage to common data structure being applied by many users is the reduction in costs of design and maintenance of application software such as modelling tools. Similarly, widely agreed and used data dictionaries clearly define the terms used and promote ease of understanding.

## When, if ever, will the cost and effort of creating the database be repaid?

This is a judgment that has to be made at the start of the project and reviewed regularly during the creation and subsequent maintenance of the database. A database is created and maintained to support the scientific needs of an individual, group or organization. If it fails to meet that need it is not cost effective. If it is doing so, but is costing the owners so much that it is undermining the scientific objectives of the group, it is also failing. Databases are always the servant and should never be the master.

The information that a database contains will usually only realize its full value within the originating organization. It may have a market value to another party, but that value is unlikely to reflect the cost of collection and maintenance. Data sets are very like used cars, they are rarely worth as much to someone else as the first owner paid for them. This is emphasized by **Miller & Gardner**. They regard data as a tradable asset that has a value determined by market forces rather than by the costs of acquisition and maintenance. Thus owners can exchange data with other organizations or sell it under license, realizing the value of a frozen asset.

## How will the database be designed and created?

Once a clear objective has been defined, the need for a digital system justified and the costs recognized, then the design decisions have to be made. For too long geological database design was simplistic. Whilst advanced methodologies of systems analysis and data modelling were developed for commercial systems (Cutts 1993) the geological sciences continued with *ad hoc* methods that were at best inefficient and brought the whole process of database design into disrepute.

An overview of analysis and design is presented by **Rasmussen**. He describes standard methods of systems analysis and database design and notes their limitations when applied to the geological sciences. **Giles & Bain** describe, in some detail, the process of analysis involved in creating a logical data model of a geological map.

One of the major problems that **Rasmussen** identifies is that geological data are not 'forms based'. These do not consist of simple fields with single values but range from text in reports through to analogue data such as seismic traces and maps.

Conventional database management systems handle such diversity of data very poorly, but technological solutions are becoming available. **Mallender** describes a system for the management of exploration data that offers one solution for the storage, retrieval and transmission of essentially descriptive information.

**Chew** also highlights some of the difficulties of modelling geological information. He notes specifically the importance of separating descriptive data from the concepts used to describe them. He also recognizes that many geological relationships that appear superficially hierarchical are in fact many-to-many.

**Coates & Harris, Doorgakant, Lhotak & Boulter, McKenzie, Power** *et al.* and **Toll & Oliver** describe, in varying detail, the design of the databases they are involved with, and in some cases present data models to illustrate their descriptions.

One aspect of database design that is rarely considered is the skill base of the potential users. This aspect of design was recognized by **McKenzie** during the design and development of the hydrogeological database for Honduras. In this case the initial design of the system and its user interface was carefully matched to the known abilities of the users. However, the system was designed to have a clear upgrade path. As the skills of the users developed with experience, new features could be added progressively.

## Where are the users?

Database systems, by their nature, are designed to provide data to their users. In most industries the users are located in comfortable computer-friendly environments like office complexes. Earth scientists are different. They belong to a very diverse profession and their place of work is the entire planet. This has a number of consequences.

One of the great advantages of digital databases is that data can readily be shared between many users on a network. But, if the users are too widely dispersed and only need to make occasional accesses to the information, the library or archive is still a valuable tool. **Bowie** describes the National Geological Records Centre (NGRC) and the creation of digital indexes to their holdings. This approach enables the users to find the information they require rapidly, without the NGRC having to bear the cost of creating and maintaining a full digital database. Where the data has to be digital, then distribution of copies of the data is a useful approach for a dispersed industry. **Allen** describes the advantages of storing digital datasets on CD-ROMs. The improved functionality of the Internet in recent years makes distributing data and information via this medium much more viable.

## Who will be responsible for maintaining the integrity of the database?

Data managers are an essential element of every database. The single most important reason for the failure of databases is the absence of a data manager. For small single user databases the owner acts as the data manager and has an intimate knowledge of the data. However, once the database becomes multi-user, a dedicated data manager becomes essential.

    **Lowe** notes that the 'traditional' geological data manager is a non-specialist who is both under-supported and under-valued. He then discusses in some detail Feineman's (1992) eight dimensions of data management, completeness, accuracy, fidelity, quality, lineage, timeliness, security and accessibility. **Lowe** goes on to argue that the data manager's role is critical especially in ensuring data quality. This theme is picked up by **Duller**, who describes the road to quality. He identifies the need for clear data management strategies. **Lowry & Cramer** and **Lowry & Loch** provide detailed descriptions of the process of data management and of specialist tools developed for the job. All these contributors stress the importance of experienced, professional staff being involved at all stages.

## Conclusion

The creation and maintenance of a successful geological database is significantly improved by careful thought and planning. A clear purpose must be defined against which success or failure can be measured. Databases need not be digital; analogue systems still have an important role to play in research. Costs, of all forms, must be carefully considered and quantified. A clear estimate of how a system will repay its creation

and maintenance must be made. Many advances have been made in systems analysis and design in the computing profession, geologists should learn from this wealth of experience. Large centralized databases are commonly inappropriate for the small, fragmented and widely distributed geological community. Finally, and possibly most importantly, data requires management by experienced, dedicated, well motivated, professional staff.

## Organization of the volume

The papers in this volume are organized into three broad sections. They are database design, data management and case studies. There is a certain amount of overlap between these headings for some individual papers but the dominant theme from each paper identified its place in the volume.

This volume arises from two meetings organized by the Geoscience Information Group, a Group affiliated to both the Geological Society and the International Association of Mathematical Geology. The meetings entitled 'Geoscience Database Design' and 'Geoscience Data Management' were held at the Geological Society Apartments on the 9th November 1992 and 8th November 1993, respectively. I thank all those who contributed and participated at the meetings and the discussion that followed. I also thank the contributors to this volume, the reviewers and the staff of the Geological Society Publishing House.

This paper is published with the permission of the Director, British Geological Survey (NERC).

## References

CUTTS, G. 1993. *Structured Systems Analysis and Design Methodology*. Alfred Waller Limited, Henley-on-Thames.

DATE, C. J. 1986. *Database Systems*. Addison-Wesley, Reading, Ma.

ELMASRI, R. & NAVATHE, S. B. 1989. *Fundamentals of Database Systems*. Benjamin Cummings, Redwood City, California.

FEINEMAN, D. R. 1992. Data Management: Yesterday, Today and Tomorrow. Presentation to the PETEX '92 Conference, 25th November 1992, London.

# An overview of database analysis and design for geological systems

KEN RASMUSSEN

*Logica UK Ltd, 68 Newman Street, London W1A 4SE, UK*

**Abstract:** Techniques for database analysis and design are well established for the development of administrative systems. When the techniques are used to develop geological databases, however, a number of difficulties arise. This paper provides an introduction to the nature of these problems. The context for the discussion is first set by describing the standard techniques for database analysis and design. The difficulties that arise when the techniques are applied to geological data are then reviewed. The paper concludes with a discussion of some recent developments in database analysis and design which will assist designers of geological systems.

Database analysis and design are now well-established techniques. They have been used for many years for the development of administrative DP systems. When they are applied to the development of geological systems, however, a number of difficulties arise. Geological data are different from the types of data found in most data processing systems and for which the standard techniques for database analysis and design were developed. Designers of geological databases need to have an understanding of these differences and the types of problems caused by using the standard techniques with geological data. They will then be in a better position to refine the standard techniques to suit the nature of their data. This paper provides an introduction to these problems and discusses some of the recent developments which may assist the designers of geological databases.

The paper begins by describing database analysis and design in general terms by answering the following questions:

- Why database?
- What are database analysis and design?
- Why are they necessary?
- What do they seek to achieve?

It then discusses some of the limitations of current methods when applied to the analysis and design of geological databases and finally reviews some of the newer developments in the field which may help designers of geological systems.

## Principles of the database approach

In this section the basic motivations of the database approach are reviewed. This allows the problems encountered by the designers of geological databases to be placed in context.

### The problem

Early computer systems were concerned with routine data processing. They involved simple predictable and repetitive tasks usually performed frequently and on large volumes of data. Such systems were highly tuned to meet their particular processing needs but as a result they were often inflexible. Over time they became difficult to change and it was therefore difficult to get them to present the information they contained in new ways. Moreover, the types of information that such systems can provide is limited.

Most information processing, however, is not simply data processing. Information needs are often not predictable. Systems are usually specified to solve particular problems, but a new problem may need different information or it may need the information to be presented in a different way. That is, information requirements are often varied to suit the problem at hand.

This is particularly true of geological systems. Geologists are constantly solving new probems. This frequently involves a re-examination and re-interpretation of old data in the light of new theories or new knowledge. Thus, geologists are re-using old data in new ways so they need a varied pool of data which can be analysed in many, varied ways.

### The database solution

The database approach was developed to address this type of more general information processing. It incorporates two key ideas:

- a database represents information by modelling reality; that is, a database is a picture of a part of reality, and as new facts are discovered they are added to the database to keep it up-to-date;

From Giles, J. R. A. (ed.) 1995, *Geological Data Management*,
Geological Society Special Publication No 97, pp. 5–11.

- the database management system used to manage the database must have a flexible query system to allow people to view the database in the various ways they require.

So, in theory, a database should reflect the structure of the information contained within it, rather than the way in which data are to be processed by a single application. In this way it can handle a wide range of processing needs with equal efficiency.

## What are database analysis and design?

In order to build a database, it is necessary to do two things. First, if we are to build a model of (part of) reality in a database then we must have a clear description of that reality. The development of this description is the task of database analysis or, as it is more usually known, data modelling.

Second, we must be able to map the data model onto a computer data structure in order to manage, process and display the model effectively. This transformation process (and in particular the modifications needed to make the processing efficient) is known as database design.

Each of these is now described in more depth.

### Database analysis (data modelling)

Database analysis, or data modelling, aims to build a model of the part of reality we wish to represent in a database. A data model has two components:

- a set of information types; these are the kinds of of facts that we want to store in the database, e.g. the rock types that are contained in a given formation;
- a set of constraints which help to ensure that the information stored in a database is sensible: constraints help to maintain the integrity of a database, i.e. the constraints try to ensure that the database will correspond to a possible reality. Whether this records the actual world depends on the facts that are entered into it. Thus the constraints cannot ensure that the database is correct. They merely help to prevent nonsense being entered. For example, a typical constraint might state that a sample can be taken from *only one* geographical location.

### The entity-relationship approach to data modelling

Many different approaches have been suggested for data modelling but by far the most widely used is the entity-relationship (ER) approach. The information types in the ER model are defined using three concepts:

- entities (things) about which information is expressed;
- relationships which are meaningful associations between entities;
- attributes or properties of entities which express information about entities.

Each of these can be grouped into classes or types. The information types are then defined by specifying the entity types, relationship types and attribute types needed to describe the information requirements of the area under study.

There are two types of explicit constraints which affect the relationship types:

- degree, which constrains the numbers of entities (entity type occurrences) that can take part in a relationship;
- optionality, which relates the existence of relationships to the creation of entities,

For example, the fact that a borehole can only be drilled at *one* position is a constraint of degree; the fact that a borehole *must* be drilled at some geographical position is a constraint of optionality.

The model also imposes some implicit constraints such as:

- all attributes must be single valued e.g. an outcrop must be of a single rock type;
- all entities must be uniquely identifiable within the model in the same way and for all time. Many entity types are given 'designed' attributes such as serial numbers to ensure that this is true. For example, attributes such as sample numbers and lithology codes normally serve solely as identifiers of concepts. If, however, these identifiers are changed over time then the model will assume that new entities are being represented because it has no way of correlating the old and new identifiers.

In many versions of the model it is also assumed that an entity can only be classified in one way, i.e. each entity can only be in one entity type. Sometimes this restriction is relaxed a little to allow entity subtypes, i.e. to allow an entity type to be a subset of another entity type. For example, both 'borehole' and 'vertical section' may be defined as entity types with boreholes being special cases (i.e. entity subtypes) of vertical sections. In such cases a single entity may belong to more than one entity type but this is usually the only situation where the restriction is waived.

*Database design*

The development of a data model is a crucial stage in the building of a database. The data model determines the types of information that will be stored in the database and ensures that the database is, at least to some extent, consistent, i.e. that it describes a possible reality.

However, if we are to use a database effectively, it must be possible to access the data readily, easily and promptly. The data model represents an ideal for providing maximum flexibility but a direct implementation using today's technology is likely to prove inefficient. It may be too slow, use too many resources (computer or human), or be too costly.

Database design aims to overcome these problems without compromising the data model too much. Unlike database analysis, database design is very dependent upon the computer environment and the DBMS used to implement the database. The database designer needs to make use of features of the DBMS (such as indexing) to ensure adequate performance. In some cases it will be necessary to alter the data model but if possible such changes should be avoided. At best they will cause an additional processing overhead; at worst they will reduce flexibility because the database is being changed to reflect particular processing needs, so defeating one of the main objectives of the database approach. However, with today's DBMSs it is usually essential to make some such changes. The skill of the database designer lies in minimizing the impact of this type of design decision.

## Problems with designing geological databases

Data modelling was developed to assist the building of databases for administrative systems. Application of the same techniques to other application areas such as geology causes difficulties. (Many of the difficulties apply to administrative systems too, but they seem to be more acute in the case of geological systems.) The problems fall into two categories:

- those caused by problems with data modelling;
- those caused by the DBMSs.

*Problems with data modelling techniques*

The types of problems caused by the data modelling techniques include:

- conflicting objectives of the modelling methods;
- inability to represent differences among entity types;
- problems with classification;
- difficulties of modelling non-factual information.

*Conflicting objectives of data modelling methods.* One of the main problems with existing methods is that data models are used to satisfy two conflicting objectives. A data model must:

- describe the reality to be represented in the database, and
- suggest data structures suitable for implementation in a DBMS.

A single method cannot satisfy both objectives. Methods which are good at describing reality require considerable translation to produce an effective database design and much information is usually lost in the translation. (It needs to be picked up in the processing.) Methods which are easily translated into current file structures (such as ER methods) are poor at fully describing the complexities of geological data. Ideally, two different models are required, one to describe our perception of reality and another to represent reality in a computer oriented (record oriented) form (see Fig. 1). In practice one model is used to serve both needs and since the ultimate goal is the implementation of a database, most people take a pragmatic approach and choose a method that can easily be translated into a database design. This effectively entails using a database design method for data analysis. The result is a struggle to model the reality effectively.

If the alternative approach of using a richer data modelling technique is used, then the designer will require much greater effort to translate the model into a design. It will also be necessary to rely on the programmers to implement much of the information from the

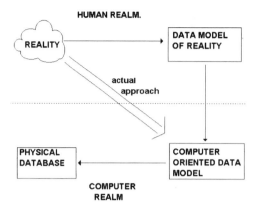

**Fig. 1.** Comparison of ideal and actual data models.

model in the program code, though the extent of this will depend on the ability of the chosen DBMS to represent constraints.

*Inability to represent differences among entity types.* Most data modelling approaches treat all entity types alike. They are just sets of things. The limitations of this assumption are very apparent in geological systems. For example, one of the primary methods of representing geological information is the geological map. The entities on a map naturally fall into different categories such as:

- point features, such as boreholes;
- line features, such as faults;
- areas, such as rock bodies.

More generally, geological systems must also be capable of representing three-dimensional objects such as surfaces and volumes.

The entities in any one of these categories have certain characteristics in common, yet our modelling methods know nothing of this. They lack even simple properties and functions that are implied by such concepts. These must be specified outside the data model. For example, the models do not include concepts such as continuity of a line and cannot enforce the constraint that areas must be closed. Moreover, there are no functions such as distance between points or direction of a line within the model.

*Problems of classification.* Classification is a fundamental technique within geology. There is a mirror to the process of classification in data modelling, namely the problem of defining entity types. For example, samples are collected for a wide variety of purposes such as for geological

study, chemical analysis or the sample may be a fossil, so we could define a single entity type, sample, or separate entity types for geological sample, chemical sample and fossil, though they would have a number of similar attributes.

The classification processes available within conventional data modelling techniques are limited. Some models support entity subtyping but this is largely just a descriptive technique. However, important concepts, such as inheritance, whereby attributes of a supertype are also attributes of the subtype, are usually missing. Further, subtyping is included in the models only as a constraint. That is, the information that, say, entity type 'fossil' is a subtype of 'sample', is defined in the data model itself but the information is not available to the user. Moreover, some important models, such as the relational model, do not even support subtypes.

This is very different from the way in which we perceive the world. The way in which we classify an object depends on what we want to say about it. For example, a piece of rock may be a fossil, a sample or a sandstone, depending on the information to be expressed. Yet if we classify entities in ways that cannot be related through a supertype/subtype association, the model will know nothing of it.

*Difficulties of modelling non-factual information.* Data modelling approaches can be compared with languages. Both have the same objective, namely to express information. However, most data modelling methods are still at an elementary stage in their ability to express information. That is, although the concepts may be more complex, the information structures that they can represent are similar. For example, although the concepts are more complex, a sentence such as 'Janet likes John', has the same structure as 'The Jurassic Period lasted 55 million years'.

Data models can only store facts. Yet, much geological information is not factual. Geologists make interpretations. A geological map, for example, will start with a set of observations and sketches made in the field and will then go through a series of refinements until it becomes the final product. At each stage of refinement the geologist will be making interpretations about the geology based on the information at his or her disposal. Yet our modelling methods can only handle interpretations as if they are facts and cannot effectively relate the different interpretations of the same facts to each other nor store the rules used to make the interpretations.

## Problems caused by the limitations of the DBMS

*Handling non-forms-based data.* Most DBMSs are designed to handle simple fields with single values. They are very good at recording such things as order numbers, part descriptions and invoice values. Some geological data can be put into this form but many data are far more complex, ranging from text in reports through to analogue data such as seismic traces and maps (though these can be digitized of course).

Present database systems and database methods are not suited to the handling of such diverse data. They have their roots in the handling of simple factual data used in data processing. At worst, such data cannot be handled at all by the DBMS. At best, text can be treated as an amorphous, unanalysed blob of data. That is, it is stored, manipulated and maintained as a single field but the internal structure is unknown to the DBMS. Digitized data can be handled more naturally but the types of data structure needed to support the data are usually absent. For example, many DBMSs would implement vector data as a set of discrete points instead of a single continuous object. There is no knowledge of such things as distance or direction within the database itself. Rather, the system designer has to implement the data with the facilities provided and write the software and constructs needed to implement such concepts. They are not provided by the data model or the DBMS. (To a large extent this problem also reflects the weakness in current data modelling techniques because it is not possible to 'picture' such information.)

*Limitations of transaction management.* For use in geological systems, DBMSs also need to change their approach to transaction management. In current DBMSs the concept of a transaction is key. A transaction is the basic unit of processing and recovery. In administrative systems transactions are usually small and fairly simple. For example, a transaction may involve the entry of the details of an order, a task which takes only a few minutes. With geological databases tasks may be much more lengthy, sometimes spreading over hours or even days. (Think of creating a geological map, for example.) To handle such activities the DBMS will need a new approach to transaction management. This will almost certainly require some form of version control built into the DBMS (not just into the applications) in order to avoid locking out users when data are being updated over a long period.

## Current trends

Although the problems described above are very real for today's designers, there are a number of developments that will help the designers of tomorrow's geological databases.

### Improved data modelling techniques

Most widely used data modelling techniques are 20 years old. Since they were developed there has been a great deal of research into data modelling and, more importantly, a growing awareness of the importance of improving the semantics of data models. A number of improvements have been proposed.

First, there should be improved support for constraints. Newer models are supported by powerful constraint languages which allow the representation of a much wider range of constraints. These languages are often based on a form of first-order predicate logic which can support all the constraints of the ER model and many more. Indeed, such a logic has been called the ideal data modelling language.

Second, the models should support richer semantics including a recognition of different kinds of entity types. Some methods can now distinguish, for example, between:

- kernel entity types, which define the main object types of interest;
- associative entity types which represent relationships between entities;
- characteristic entity types which qualify or describe another type of entity.

Third, a breaking down of some artificial or unclear and unnecessary distinctions such as the differentiation between attributes and relationships and the differentitation between entity and entity type. Is the location where an outcrop occurs a property of the outcrop (an attribute) or is position an object in its own right. There is no clear answer. Some methods resolve the dilemma by not having attributes at all; just entities and relationships. Similarly, if a concept such as a borehole is defined as an entity type we cannot express the fact that a particular object is a borehole. That is, the database will not record the entity type to which an object belongs in a form that makes it accessible to a user. Indeed, the classification must be known before programs which access the data can be written. This is because the concept of a borehole is an entity type (not an entity) and therefore puts it at a higher level. It is defined in another database, the data dictionary. Such distinctions are unnecessary from an information modelling viewpoint. They are imposed by the limitations of current DBMSs. There have been attempts to remove such restrictions in both theoretical models and in some practical applications. Such applications have usually been developed under the banner of artificial intelligence but as AI and conventional data procesing have moved closer together I believe that we shall see a greater integration between the two.

### Object-oriented approaches

Object-oriented models are good examples of the more recent approaches to data models. Object-oriented models are a family of models rather than being a single model but they share a number of common characteristics.

- Greatly improved support for entity subtyping: in particular, inheritance of properties is a key feature. For example, if 'fossil' and 'geochemical sample' are defined as subtypes of 'sample', they can inherit attributes of 'sample' such as the position where the sample was found and date on which it was found (see Fig. 2).
- The recognition that the behaviour of an object is a property of that object. In Object-oriented techniques this means associating processing with an object. It provides an effective means of representing constraints.
- Improvements in the semantics of relationship types. A range of different kinds of relationship types can be defined. For example, if we were able to distinguish 'part of' relationships, which express the fact that one object is a constituent part of another, then constraints such as the fact that a

formation cannot have a greater geographical extent than the group of which it is a part, would be enforced naturally.

### Improved DBMSs

It is relatively easy to develop a new data modelling approach, but that is only half of the answer. Improved modelling methods need to be supported by a DBMS capable of implementing the constructs on which they are based. However, it is much harder to develop the software for a new DBMS than to develop a new modelling method. Indeed, the biggest stumbling block to the wider use of the newer semantic data models has been the lack of a supporting DBMS. However, some progress is being made towards the development of more sophisticated DBMSs. Again the object-oriented models are leading the way. We are already beginning to see DBMSs in which:

- constraints are enforced by the DBMS itself rather than by the application software;
- new data types, and more importantly, appropriate functions on those data types, can be defined within the DBMS.

Both are examples of the gradual evolution of conventional DBMS towards an object-oriented DBMS.

### Database languages

In addition to changes in the DBMS, we are beginning to see corresponding changes to the database languages used by the DBMS. In particular, the database language SQL is being enhanced to provide a number of the features described above. For example, there are proposals to introduce new data types and functions particularly for handling spatial data into the language. Many of these are of direct relevance to geological databases, e.g. data types such as point, line, polygon. These new data types may be complemented by new predicates such as 'near to', 'encloses' and 'to the west of'.

### Storing text and other objects

We have already seen that one of the major problems of geological data is that much of the data is not 'forms based'. It cannot, therefore, be easily stored or managed by conventional databases. However, this is beginning to change. Improvements in technology (especially

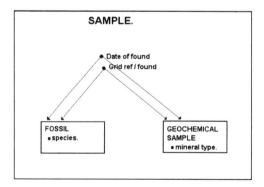

**Fig. 2.** Inheritance of attributes by subtypes.

those leading to falling memory prices) mean that it is now possible to store images and text as single fields within a database and to manipulate them using database languages. This offers great potential for developing systems in which forms-based data is closely integrated with the text and images from which it is derived.

## Conclusions

I once read a mathematical paper entitled 'A surrealist landscape with figures'. The field was claimed to be largely unknown territory but there were a few significant results. I believe that the use of databases in geology is similar. Although the techniques we have are still in their infancy, some significant progress has been made. Yet there is still a long way to go. We can see some useful advances in the near future. We can also see that it is subjects like geology, with the extra demands that they make on database analysis and design, that are going to help shape and test the future of database technology.

# Data modelling a general-purpose petroleum geological database

## K. J. CHEW

*Petroconsultants SA, PO Box 152, 24 Chemin de la Mairie, 1258 Perly, Geneva, Switzerland*

**Abstract:** Experience gained while modelling a petroleum geological database can be summarized as follows. General-purpose database models should reflect the real world, complete with imperfections, and avoid being application-oriented in design. In geological database modelling it is necessary to separate the recording of descriptive features (empirical data) from that of geological processes (concepts). Detailed analysis reveals that many geological relationships assumed to be inherently hierarchical are in fact many-to-many. Ambiguity and inconsistency in definitions should be avoided, if necessary by adopting a definition at variance with industry norms. Use of generic entities simplifies the modelling exercise and can benefit the end-user.

In this paper the emphasis is on the 'general-purpose' nature of the data-modelling exercise for a 'corporate' or 'reference' database, the type of database in which are stored basic factual data and the results derived from 'interpretation' or 'project' databases.

A large number of highly efficient project databases undoubtedly exist. In general, however, these cover a relatively small area (a field, basin or even country), and have been developed with specific types of business application in mind by a company or organization which has control over both the availability of data and the quality of available data.

Experience suggests that problems are liable to arise with such a database when an application significantly different from those envisaged at the design stage is required, or when it becomes necessary to incorporate entities and attributes not considered at the design stage.

It follows that a model for a general-purpose database should not be application-oriented in design. (It should be noted that for the purposes of this paper, the logical data model, its implementation as a physical database and the software required to load datasets and maintain the data within them are not considered to be business 'applications'; they represent the kernel of the system.)

While a truly application-independent model is an ideal, it is probably an unattainable one. Nobody produces a data model and creates a database for no reason; behind each model lies the designer's view of the data and behind each database lies the need for its existence.

## The Model 'world'

Whatever the view of the data modeller, account must be taken of the real world, with all its imperfections, which the data model sets out to describe and summarize and from which the data to populate the database will come. If a model does not approach reality (and reality includes recognizing that almost any item of key data may not be available), the benefits of the powerful organization of data will be lost as a result of the exclusion of data that do not fit the model or the inaccurate portrayal of data within the model.

When Petroconsultants commenced the design of the IRIS21 Logical Data Model, an 'ideal' world was assumed, which was subsequently modified to allow for representation of data as we really receive it. The ideal world presumed that most items in a petroleum exploration database can be fully described in four dimensions (space and time).

Thus, if one knows the well, date and depth from which a fluid sample was obtained and the dates and intervals over which a series of well tests were conducted, it should be possible, without further description, to associate the sample with a specific test. In theory, even depths are not required to associate the sample with the test if the times of testing and collection have been accurately recorded (Fig. 1).

In practice, data gaps require that the model contains sequence numbers for many records, both to record the sequence in which events occurred and to associate items (such as tests to samples).

Similarly, in lithostratigraphical description, accurate dating should allow one to establish the temporal relationship between two units. In practice, we will probably have to introduce a series of temporal qualifiers (before, overlapping, equivalent, etc.) as proposed by Davis (1992) to describe relationships such as those between formations, whether faulting occurred after hydrocarbon migration, etc. Even greater

*From* Giles, J. R. A. (ed.) 1995, *Geological Data Management*, Geological Society Special Publication No 97, pp. 13–23.

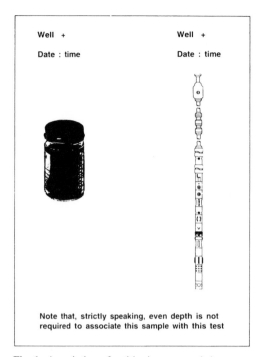

Well +

Date : time

Well +

Date : time

Note that, strictly speaking, even depth is not required to associate this sample with this test

**Fig. 1.** Association of entities in space and time.

problems arise in describing spatial relationships, such as whether faulting penetrates the reservoir interval. Again, qualitative description may provide a more easily computed answer.

## The geological 'world'

The US General Accounting Office recently observed 'that scientific databases are different in nature to business databases' (Gavaghan 1992). There are probably some cynics amongst us who would suggest that geological databases are different from scientific databases!

### What is different about geological databases?

It is worth pointing out to non-geologists involved in modelling geological data, that the geological data modeller faces particular problems because of the nature of geological information and the manner in which geological science is practised.

### Geological information

Geological databases resemble other scientific databases in that they contain the results of measurements and observations. What is distinctive about them, however, is that geological description is four-dimensional and involves being able to describe variation in the shape, relationships and distribution of spatial objects through time.

Because three-dimensional geological objects (such as a reservoir sandstone body, a fault plane or an unconformity surface) can rarely be observed and measured in their entirety but have a geometry and topology which must be inferred, geological description is frequently qualitative or at best semi-quantitative and is the result of the interpretation of an individual or group.

Similarly, the history of geological objects through time is rarely quantifiable but on the contrary is normally an interpretation based on an individual's estimation of the effect of geological processes over the relevant time interval.

Such interpretations are made on the basis of geological procedures and practices that have evolved over the past two centuries and take little account of the requirements of modern databases!

### Geological procedures and practice

Geology is extremely diverse in scope: the comprehensive study of geological phenomena requires the application of chemistry, physics, mathematics and the life sciences. Unlike the majority of these sciences, geology has no system of natural laws and relatively few guiding principles which meet with worldwide acceptance. Developmentally, it therefore remained under the influence of philosophy and theology for much longer than other sciences.

To this day, geology employs the 'mental procedure' of multiple working hypotheses expounded by Chamberlin at the end of the nineteenth century (Robinson 1966), i.e. proposing and, where possible, testing several tenable hypotheses against observed phenomena in an impartial manner. Most of these hypotheses are too complex to express mathematically. For this reason, geology remains largely a descriptive science.

Because it is descriptive, geology suffers from the many problems that arise from descriptive schemes: there is a lack of globally accepted classification schemes; long-established classification schemes often have little logical basis; new/refined classification schemes are frequently being created; and many descriptive classifications have genetic implications embedded within them. In many cases, these in turn reflect the fact that much geological practice has strong political and cultural ties. (As an example, Petroconsultants has identified 20 classes of lithology (e.g. mudrock, carbonate, volcanic, hornfels) which are sufficiently different in their accepted descriptive scheme to warrant an individual descriptive scheme.)

One consequence of this situation is that any worthwhile attempt to model the geological world will almost inevitably involve the rationalization (and normalization) of existing classification schemes if it is to provide a basis for geological description in the consistent manner that is required. While this may be no bad thing if it results in more rigorous geological thinking and less ambiguity in geological description, it is not without its dangers! Different geologists (and companies) will wish to use different schemes and it must always be possible to populate the database with data from older, outmoded descriptions. We will see below how database design can itself assist in fulfilling these objectives.

## Description versus interpretation

In the case of a petroleum geological database, we start from the premise that we wish to record the geological information acquired in the course of hydrocarbon exploration, the geological characteristics of discovered hydrocarbon deposits and the geological environment in which these hydrocarbon accumulations occur.

As we move from the geological results of a single well, through the description of an accumulation, to the petroleum geological context in which an accumulation occurs, our sampling interval decreases exponentially and the interpretive element in our description increases accordingly.

At the well level it is normally possible to record a detailed description of observed features. As the amount of interpretation required increases, we find built into the recorded observations the results of the geological processes which the interpreter believes to have taken place. In other words, there is a progressive shift from the empirical to the conceptual in what is described.

### Data types

The second point to stress in geological modelling is therefore the need to separate the recording of descriptive features from that of geological processes. Empirical data are used to

develop, test, refine and even dismiss concepts; the two types of observations must be kept distinct.

On the subject of empirical data, it is also important to note that they come in several types: *raw* (or measured) data; data to which some correction or conversion function have been applied; and data which have been computed from other data values. The latter two types fall into the category of *processed* data.

Note that these empirical data types require to be distinguished from one another within the physical database and the facility to do so must therefore be incorporated into the model. For example, if a correction or conversion factor used within a database is subsequently found to be in error or is refined, it will be necessary to adjust *all* stored values previously based upon that factor. (An example might be the reconversion of Danish annual gas production from cubic metres to standard cubic feet after discovering that it had originally been reported in normal cubic metres, not standard cubic metres.)

In the case of a computed (derived) value (such as a recoverable oil reserve estimate), it is necessary to be aware that it is computed so that the value can be adjusted if any of the parameters from which it is derived are modified (such as porosity or oil saturation in the case of oil reserves).

### Data quality

All types of geological data are potentially of variable quality. Allowance should be made for this by the possibility to report extreme ranges (minima and maxima), percentage accuracy, qualitative values (low sulphur, heavy oil) and the complete absence of data. Variable-quality data result from the reliability or otherwise of the origin of the data, hence it is also essential to report the data source to allow the user to assess and weight data according to their perceived dependability.

In addition, both processed empirical data and all interpretive data are associated with an interpreter and a geological data model should enable this aspect to be reported. In fact, Petroconsultants' IRIS21 Logical Data Model allows the same attribute to be reported multiple times if it is associated with a different source *or* a different interpretation.

### Metadata

The problem of describing data by means of alternative or redundant classification schemes has already been mentioned. A solution to problems of this type is just one of several areas in which metadata (data about data) can be used to enhance the quality, versatility and reporting capability of a database.

Databases can be designed to store a number of useful pieces of information about data such as:

- definition: records any quantifiable definition of the reported value, e.g. 'silt' has a size scale of '0.004–0.062 mm' in size class;
- classification: records whose definition or classification the value belongs to, e.g. 'modified Wentworth size scale';
- synonym: records which values are equivalent to other values, e.g. particle form 'discoidal' (Folk) = 'oblate' (Zingg) = 'platy' (Sneed & Folk);
- qualifier (adjectival form): useful in context-dependent reporting, e.g. when reporting component qualifier of sedimentary rock, use adjectival form 'silty', not noun 'silt';
- validity: records where any particular value may legitimately be used, e.g. the size qualifier 'silty' applies only to the lithology classes 'sandstone' and 'conglomerate'.

## The real nature of geological relationships

The shift from the empirical to the conceptual as we describe progressively larger-scale features is paralleled in the description of geological entities (such as a 'megasequence') and petroleum-geological entities (such as a 'play') (Table 1).

Perhaps because geologists in many branches of the science apply systematics, there has been a tendency when modelling the above data entities to assume that a hierarchical relationship exists and that the petroleum-geological entities can be dovetailed with the geological ones in this hierarchical classification. This is not the case.

With one or two possible exceptions, the relationships between the eight entities listed in Table 1 are many-to-many (Table 2). As an example of a many-to-many relationship, a well may encounter more than one lithology (sandstone, limestone); a lithology (sandstone) may occur in more than one well.

Not all of the many-to-many relationships indicated in Table 2 are as obvious as in the above example. For example, it might seem that an accumulation should only occur in one play. But one can visualize the situation in which a reservoir in a supra-unconformity fault scarp talus play will, on some occasions, be contiguous with a reservoir in a sub-unconformity tilted fault block play, forming a single accumulation. Is this an example of an accumulation occurring in two plays or does the combination of a supra-unconformity fault scarp talus play with a sub-unconformity tilted fault block play represent a third type of play? It depends entirely on how one defines 'play'.

Similarly while one might imagine that a single megasequence or hydrocarbon accumulation should be restricted to one geological province, this need not always be the case, especially where provinces are superimposed or if the limits of the geological province have been chosen on the basis of defective criteria.

## Dictionaries and definitions

This brings us to another problem area, that of definitions. One cannot simply pick up the *AGI Glossary of Geology*, the *Petroleum Industry Data Dictionary*, the *Public Petroleum Data Model*, a classic textbook or a seminal paper and hope to find a definition that can be plugged

## Table 1

The Real Nature of Geological Relationships (Parallelism of Relationships)

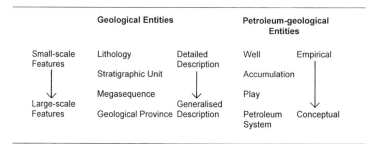

| | Geological Entities | | Petroleum-geological Entities | |
|---|---|---|---|---|
| Small-scale Features | Lithology | Detailed Description | Well | Empirical |
| | Stratigraphic Unit | | Accumulation | |
| | Megasequence | | Play | |
| Large-scale Features | Geological Province | Generalised Description | Petroleum System | Conceptual |

## Table 2

The Real Nature of Geological Relationships (or Many Many-to-Many)

|               | Lithology | Strat Un | Megaseq | Geol Prov | Well  | Accum | Play  | Petrol Sys |
|---------------|-----------|----------|---------|-----------|-------|-------|-------|------------|
| Petrol System | M - M     | M - M    | M - M   | M - M     | M - M | M - M | M - M |            |
| Play          | M - M     | M - M    | M - M   | M - M     | M - M | ? - M |       |            |
| Accumulation  | M - M     | M - M    | M - M   | M - M     | M - M |       |       |            |
| Well          | M - M     | M - M    | M - M   | M - M     |       |       |       |            |
| Geol Prov     | M - M     | M - M    | M - M   |           |       |       |       |            |
| Megasequence  | M - M     | ? - M    |         |           |       |       |       |            |
| Strat Unit    | M - M     |          |         |           |       |       |       |            |
| Lithology     |           |          |         |           |       |       |       |            |

directly into a data model. Definitions must conform to the view of geological processes that one is attempting to describe through the model and to do this they must satisfy a number of criteria.

### Elimination of ambiguity

Consider the notion of the 'play', a concept intrinsic to modern hydrocarbon exploration. Literature definitions range from the highly detailed to the vague ('activities to explore for hydrocarbons') and from the purely geological to those which include the activities involved in exploration (Table 3).

The Petroconsultants definition of 'play' is as follows: an association of proven and/or prospective hydrocarbon accumulations at a specific stratigraphic level, which share similar geological conditions of entrapment of hydrocarbons such as lithology and depositional environment of reservoir and seal, and trap type.

This definition differs from that of Bois (1975), which requires hydrocarbons to be of similar composition; that of White (1988), which requires a similar source; and that of Allen & Allen (1990), which requires a common petroleum charge system but not a similar trap. The definition of Miller (1982), on the other hand, is compatible with that of Petroconsultants. It is important to note that Petroconsultants do not consider the definition used by them to be more 'correct' than the others, but simply that it is consistent with the hydrocarbon generation and accumulation process as it is modelled within Petroconsultants' IRIS21 system.

It is clear that in the appropriate circumstances (such as the Moray Firth Basin example given in Fig. 2), a population of hydrocarbon accumulations could be subdivided into four separate (but overlapping) groups, based on the application of the four different definitions outlined in Table 3. The definition of Miller, if applied even-handedly, should yield similar results to that of Petroconsultants.

**Table 3.** *Definitions of 'Play'*

*Bois (1975)*
'A continuous portion of sedimentary volume which contains pools showing the following characteristics:
(1) reservoirs within the same productive sequence occur throughout the zone
(2) hydrocarbons are of similar chemical composition, and
(3) traps are of the same type.'

*Miller (1982)*
'A practical meaningful planning unit around which an integrated exploration program can be constructed. A play has geographic and stratigraphic units and is confined to a formation or group of closely-related formations on the basis of lithology, depositional environment or structural history.'

*White (1988)*
'A play is a group of geologically-related prospects having similar conditions of source, reservoir and trap.'

*Allen & Allen (1990)*
'A family of undrilled prospects and discovered pools of petroleum that are believed to share a common gross reservoir, regional top-seal and petroleum charge system.'

*Petroconsultants (1992)*
'A play is an association of proven and/or prospective hydrocarbon accumulations at a specific stratigraphic level, which share similar geologic conditions of entrapment of hydrocarbons such as lithology and depositional environment of reservoir and seal, and trap type.'

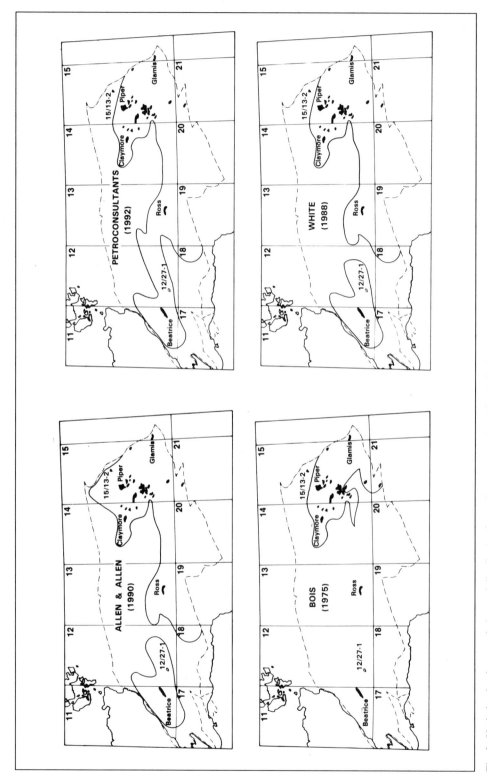

**Fig. 2.** Variation in play population and distribution in sandstone reservoirs of the Callovian–Tithonian paralic to shallow-marine transgressive sequence of the Moray Firth Basin, as a result of applying different definitions of 'play'.

Several other definitions of 'play' exist but they would tend to result in one of the above four groupings if applied to a family of deposits.

On the other hand, the 'play' definition of Crovelli & Balay (1986) is 'a collection of prospects having a relatively homogeneous geologic setting'. This definition embraces all of the above four definitions, which as we have seen can give rise to four different results. It is therefore too non-specific to be of value in a data-modelling and database context.

Whatever definition one chooses, one test of the validity of a definition of an entity (especially a conceptual one such as 'play') is that it should leave as little ambiguity as possible when it comes to assigning a member of a population to a specific class.

## Coherence of definitions

A second problem associated with definitions in geological data modelling occurs when entities, which by their nature are supposed to be related to one another, have definitions which lack 'fit', i.e. their definitions either overlap or leave a critical gap.

This can be illustrated by an example involving the description of source rock characteristics within a petroleum-geological data model. Depending on the definition one uses, one can associate a source rock with the stratigraphic unit of which it forms a part, with a play (as do White (1988) and Allen & Allen (1990)) or with a petroleum system (a cell which comprises a coherent package of mature source rock(s) and any existing and/or former primary hydrocarbon accumulations generated therefrom—Petroconsultants' definition). Where in the database should the characteristics of this source rock be described?

One aspect which should be taken into consideration in answering such a question at the database design stage is the likelihood that the entity with which source rock characteristics should be associated will exist within the physical database. Not all of them will be present all of the time. To model this effectively we have to examine the process whereby we accumulate knowledge about source rocks, i.e. the exploration process.

If we pursue the example of a source rock in an undrilled offshore basin with no seabed sampling or knowledge of the stratigraphy, the only stratigraphic data available are from seismic stratigraphy and analogy with adjacent provinces.

It is then possible to drill a well which fails to make a discovery (no proven accumulation or play) but which finds mature source rock, indicating the presence of a petroleum system. No formal lithostratigraphy would be established on the basis of a single well, and the source rock description would simply be part of the geological description of informal rock units observed in the well.

An alternative scenario is that the first well finds an accumulation, thereby confirming the existence of a play, but finds no evidence concerning the nature of the mature source rock, leaving us with nothing upon which to found a petroleum system.

Both of these scenarios are distinct possibilities in the real world and our model must be capable of describing them. To accomplish this, Petroconsultants has found it necessary to exclude similarity of source or hydrocarbon character from the definition of play. This does not mean that a play definition which incorporates these features is wrong; just that it does not fit well with the definition of petroleum system and the hydrocarbon generation and accumulation process as Petroconsultants has modelled it.

Note that because any element of sourcing is excluded from the definition of play, it is possible to have an unproven play concept in which all the prospects subsequently turn out to be dry (no source) and a petroleum generative system which is devoid of hydrocarbon accumulations (no traps), i.e. we truly model the real world!

The situation in which a proven play is sourced from a known source rock is described by the association of a play with a petroleum system and the association of that petroleum system with the stratigraphic unit from which the hydrocarbons were generated. Note that this model allows a single play to be sourced from more than one petroleum generating system, one petroleum system to source multiple plays and a common source stratigraphic unit to have different sourcing characteristics (e.g. time of peak hydrocarbon generation) within different petroleum systems—another example of the need to report many-to-many relationships.

Hopefully, the foregoing example highlights the need to tailor definitions, at least of conceptual entities, to fit the model rather than distorting the model to fit published definitions. Bear in mind that these are unlikely to have been prepared with a general-purpose data model in mind but are more likely to have originated as 'once-offs' in publications focussing on specific applications.

We can make further use of the source rock example to illustrate the different treatment of empirical and conceptual data.

*Separating fact from interpretation*

Within the IRIS21 data model, Petroconsultants has chosen to record measurable organic geochemical characteristics (kerogen type, TOC, VR, etc.) as part of the lithologic unit entity associated with a stratigraphic unit (lithostratigraphic or chronostratigraphic), i.e. at the empirical level.

On the other hand, interpreted features which describe hydrocarbon generation and which are dependent on the rock unit being proven as a source rock (e.g. maturity/post-maturity depth, area/volume of mature source, age range(s) of generation) are recorded along with the source rock description of the specific petroleum system with which they are associated, i.e. at the conceptual level.

Modelled in this way, an organic-rich lithological unit is not associated with hydrocarbons *per se* but only becomes a source rock when associated with one or more petroleum systems.

*Testing the model*

The ultimate test of the definition of the model is that the kernel of the system (as defined above) should function, i.e. it should be possible to put all necessary data into the system without misrepresenting them. The careful specification of the definitions of 'play', 'petroleum system', etc., given in the examples above were made in full knowledge of both the hydrocarbon generation and accumulation process and the real-world situations likely to arise (e.g. missing data, unknown information, unsuccessful wells).

, However, it is virtually impossible to get the model correct at the first pass. The process has to be an iterative one of designing a model, implementing it as a database and attempting to add data to that database. Where data cannot be added to the system in a manner which adequately portrays those data, the model must be re-examined and, where necessary, refined.

An example from the development of the IRIS21 data model illustrates this process. It was assumed that the relationship between a lithostratigraphic unit and its geological description would always be a direct one. In practice, it was discovered that certain widely distributed units have highly variable geological characteristics

which are related to the different geological provinces in which they occur. (The Kimmeridge Clay is a classic example, ranging as it does from Dorset to the Viking Graben.) It was therefore necessary to refine the model to allow a geological description to be linked to the association of a specific stratigraphic unit with a specific geological province.

## Recognizing 'generic' entities

A particularly useful concept in a petroleum-geological database is that of 'generic' entities. A generic entity within the IRIS21 data model is defined as an entity that can occur in association with two or more other major entities (e.g. wells, fields) but which is described in the same terms throughout the system.

If one looks at the example of hydrocarbon production one can see that one might wish to associate production with an individual tubing string, a well, a reservoir sub-unit, a reservoir, a field, an operator, a contract area, a basin, a country or some even larger geopolitical entity such as OPEC. Regardless of which of these entities one wishes to record production for, the terms in which production is described are essentially the same, i.e. production is a generic entity.

There are a large number of facets of geological description which are generic in the above sense. In Table 4, examples are given of 16 different geological attributes which may be used to describe geological entities at various scales ranging from geological provinces to individual lithologies.

*Advantages*

Generic entities are used to describe different things in the same way. Besides greatly reducing the modelling exercise, this also helps the user and software developer by providing the benefit of a common, familiar format.

With good database design, generic entities should also permit the user to describe the same thing in different ways. Take, for example, the description of the stratigraphic information obtained by drilling a well. There are numerous ways in which to record these data. The data model should allow for any of these (plus some that have not yet been thought of!).

One possibility is to describe the stratigraphic column in chronostratigraphic terms, e.g. the Upper Cretaceous extends from 1234 m BRT to 2345 m BRT and the lithology is chalk.

## Table 4

Selected Examples of Generic Description of Geological Attributes

| Descriptive Parameters | Geologic Province | Megasequence | Stratigraphic Unit | Lithology |
|---|---|---|---|---|
| Plate Tectonic Env | X | X | | |
| Volume | X | X | X | |
| Area | X | X | X | |
| Gross Thickness | X | X | X | |
| Depth | X | X | X | |
| Age/Age Range | X | X | X | |
| Geometry | X | X | X | X |
| Depositonal Envir | X | X | X | X |
| Lithology | X | X | X | X |
| Petrophysics | X | X | X | X |
| Paleo Temperature | | X | X | X |
| Sedimentary Structures | | X | X | X |
| Thermal Conductivity | | | X | X |
| Fluid Physics | | | X | X |
| Biogenic Structures | | | X | X |
| Bedding | | | X | X |

A second is to describe the column in lithologic terms: a chalk unit extends from 1234 m BRT to 2345 m BRT and its age is Late Cretaceous.

A lithostratigraphic description is a third possibility: the Chalk Group extends from 1234 m BRT to 2345 m BRT, its lithology is chalk and it is of Late Cretaceous age.

Many models allow for only one of these possibilities. Some are even more restricted in that they describe lithostratigraphy, shoe-horning the age into a range associated with the lithostratigraphic unit in a table and thereby eliminating the possibility of recording diachronism and the extent of erosion or non-deposition.

A fourth and potentially more flexible approach is to record the three types of stratigraphic information independently by depth. This allows the modification of one of the attributes without influencing the other dependencies. Using the above example of chronostratigraphical, lithological and lithostratigraphical description, if the three are described independently by depth, the chronostratigraphy can be modified to: Danian— 1234 m BRT to 1342 m BRT; Upper Cretaceous—1342 m BRT to 2345 m BRT, without having to modify the lithological or lithostratigraphical description.

Because each type of stratigraphical description is simply a value in a table, additional types of stratigraphy (e.g. biozones) can be recorded in parallel with the other data while new types of stratigraphical subdivision (e.g. stratigraphic sequences) can be introduced with a minimum of effort. Other types of geological zone or event that can be recorded in parallel with the foregoing ones include reservoir zones, seismic horizons, faults, unconformities, marker beds and data such as magnetic polarity reversals.

Possibly the most important generic entity of all is geological description. This allows any attribute of geological description to be applied to any or all of the types of geological interval or horizon that the user may wish to describe now or in the future.

Another aid to flexibility comes from categorizing 'types' of things. If all attributes that have the potential to vary are identified in this way, each variant can be recorded in the associated table. In this way, items can be added to the database in the time it takes to update a table, normally seconds. Thus a new downhole tool, a new sedimentary texture or a new basin classification can be implemented with immediate effect.

### Drawbacks

There are, however, some problems associated with generic entities. As always, a clear definition is required for each attribute. Take, for example, the attributes of depth, elevation, height and thickness. There are a number of circumstances in which geologists use these terms interchangeably. Yet they really represent two separate things. Thickness is a 'size' attribute whereas elevation is an attribute of 'location', with height and depth merely being variants of elevation (i.e. elevation associated with a (positive or negative) 'direction').

Well-designed applications software will normally present the user with the 'height/depth' terminology which he/she expects to see in any specific circumstance, while preserving the data as positive/negative elevations in the physical database.

Nevertheless, there are problems associated with 'generic' solutions, some of which have yet to be resolved within the IRIS21 system. These principally relate to context-dependency. For example, how to provide on-line help appropriate to the specific circumstances in which the generic entity is being used and, similarly, how to constrain data values where the acceptable values in a specific context are more limited than those of the entity as a whole (e.g. numbers must be positive; degrees must not exceed 90°).

*Where to draw the line*

A further issue is how far back should we push the generic description. Thickness is an attribute of size. So also are length, breadth, area and volume (and the units used to measure these). The association of these attributes goes even further in that there is also a relationship between length × breadth and area and between area × thickness and volume. Do we record all of these attributes, some of which may be related, in a single entity called 'size' or do we regard each as an entity in its own right? Since the end-user does not require to know that the thickness of a formation is an attribute of size, to some extent the decision on how to model this aspect is one for those responsible for constructing and maintaining the physical database to take.

The decision which is arrived at will inevitably be influenced by the manner in which the database is used and whether or not the creation of an additional layer of entities is likely to enhance the functionality of the system more than it degrades its performance.

**Conclusions**

We have now returned full circle to the point made at the beginning of the paper: while a truly application-independent model is an ideal, it is almost certainly an unattainable one. There is no such thing as the unique, 'correct' geological data model; the designer's view of the data and the overall purpose of the database have an inevitable influence on the model which is eventually produced.

In this paper, a number of aspects and approaches to geological data modelling have been discussed. These are based on Petroconsultants' experiences in modelling the geological section of IRIS21. Some of these insights are the result of careful analysis and forethought; others were learned the hard way!

Either way, it is felt that they form the basis of a useful checklist for those likely to be involved in geological data modelling.

- Define relationships as they exist in the 'ideal' world in which all information about an entity is available; modify to reflect the 'real' world in which almost any information may be unreported or unknown.
- Be prepared to rationalize and normalize existing classifications, but ensure that existing and historic systems can be cross-referred to the new structure.
- Separate empirical and conceptual data.
- Distinguish between raw and processed empirical data.
- Account for variable data quality by allowing ranges, percentage accuracy, qualitative values, source and interpretation.
- Use the metadata concept to record quantifiable definitions, classification, synonyms, and validity of use.
- Check for many-to-many relationships; there will be fewer hierarchical relationships than you think.
- Use the least ambiguous definition possible; remember that someone has to use your definition to judge how to populate a database.
- Be prepared to adjust definitions, especially of conceptual entities, to ensure that the model is coherent (no overlaps, no gaps) and that it continues to function in the 'real' (imperfect) world of absent information.
- Consider the use of 'generic' entities where appropriate; used with care they can reduce software development costs and increase user familiarity and 'comfort'.
- A data model can only be considered truly tested and functional when data can be loaded into a physical implementation of the data model without misrepresenting those data.

The author wishes to acknowledge the contribution of H. Stephenson, principal architect of the IRIS21 data model, who developed the geological portion of the model with the author's assistance. Thanks are also due to P. Archer, Petroconsultants, and two anonymous reviewers for their helpful comments on drafts of the manuscript.

# References

ALLEN, P. A. & ALLEN, J. R. 1990. *Basin Analysis—Principles and Applications*. Blackwell Scientific Publications.

BOIS, C. 1975. Petroleum-zone concept and the similarity analysis contribution to resource appraisal. *In: Methods of Estimating the Volume of Undiscovered Oil and Gas Resources*. AAPG Studies in Geology, **1**, 87–89.

CROVELLI, R. A. & BALAY, R. H. 1986. FASP, an analytic resource appraisal program for petroleum play analysis. *Computers and Geosciences*, **12**, 423–475.

DAVIS, W. S. 1992. Temporal and spatial reasoning about geological history. *In: Proceedings 1992 Conference on Artificial Intelligence in Petroleum Exploration and Production, Houston, Texas.* 130–137.

GAVAGHAN, H. 1992. Plans for massive NASA database 'too vague'. *New Scientist*, 7 March, p 22.

MILLER, B. M. 1982. Application of exploration play-analysis techniques to the assessment of conventional petroleum resources by the USGS. *Journal of Petroleum Technology*. **34**, 55–64.

ROBINSON, S. C. 1966. *Storage and Retrieval of Geological Data in Canada*. Geological Survey of Canada, Paper **66–43**.

WHITE, D. A. 1988. Oil and gas play maps in exploration and assessment. *AAPG Bulletin*, **72**, 944–949.

# Database design in geochemistry: BGS experience

## J. S. COATS & J. R. HARRIS[1]

*Minerals Group, British Geological Survey, Keyworth, Nottingham NG12 5GG, UK*
[1] *Present address: Nottingham City Hospital, Hucknall Road, Nottingham, UK*

**Abstract:** The Minerals Group of the British Geological Survey (BGS) has been successfully using a relational database for over seven years to hold geochemical and related locational and geological data collected by the DTI-sponsored Mineral Reconnaissance Programme (MRP) since 1972. Operational experience has shown that several problems and deficiencies of the database are the result of the lack of normalization of the original data model. A new data model, derived from analysis of the data collected by the MRP and other geochemical programmes, is presented and solves many of these problems. It is designed to hold all BGS geochemical data for the UK landmass and is currently being implemented. It is presented as a contribution to a wider debate on generalized data models for geochemistry.

All databases attempt to model reality and hold information about their world of interest. These models can vary in complexity and, also, change and evolve with time. This paper reflects the experience of the Minerals Group of British Geological Survey (BGS) in using a simple database design for the past six years and the design principles employed in creating a new conceptual schema or logical model for a geochemistry database.

The Minerals Group of BGS, or its predecessors, has been carrying out geochemical exploration surveys for the past 30 years, initially in connection with uranium exploration in the UK but, for the past 20 years, conducted under the Mineral Reconnaissance Programme (MRP). The MRP is funded by the Department of Trade and Industry with the aim of encouraging exploration in the UK by mining companies. BGS has completed 127 reconnaissance mineral exploration projects over this period and these have been published as a series of MRP reports (Colman 1990). Notable successes of the MRP have been the discovery of the Aberfeldy baryte deposits (Coats *et al.* 1980) and, more recently, gold in south Devon (Leake *et al.* 1992). The MRP has always been multidisciplinary in nature, with input from geologists and geophysicists, but this paper is chiefly concerned with geochemical and related geological data. Geophysical data collected as part of the MRP have been incorporated into an index-level geophysical database run by the Regional Geophysics Group of BGS. Geochemical work has also been carried out in several areas outside of those described in MRP reports. Unpublished internal reports may exist for the areas or, if the study was inconclusive, the data left in paper or

computer files. One of the primary aims of the MRP database project has been to make these data accessible to BGS staff and to exploration companies.

The greater use of geochemistry in mineral exploration over the 30 year period has been made possible by the increasingly automated methods of analysis, which have changed from essentially manual methods, such as colorimetric, DC-arc emission spectrography and atomic absorption spectrometry, to the rapid, multi-element X-ray fluorescence and ICP spectrometry techniques in use today. Data recording methods have also changed over this period, with the early results being recorded on paper, then on pre-printed results sheets for data entry on punched cards and, finally, produced by computers incorporated into the analytical equipment and communicated over a site network or communication link.

The quantity of geochemical data collected by the MRP and related projects is estimated to be about $1.6 \times 10^6$ analyses by various methods on 125 000 samples. An accurate estimate is difficult to calculate because of the lack of a complete index to the number of analysed samples but records of the BGS analytical laboratories allow this approximate estimate to be made.

## Real system

The real system modelled by the database is specific to geochemical mineral exploration but this area of interest is not very different from geochemical mapping or lithogeochemistry. These methods are described in textbooks, for example the volumes comprising the *Handbook*

*From* Giles, J. R. A. (ed.) 1995, *Geological Data Management*,
Geological Society Special Publication No 97, pp. 25–32.

*of Exploration Geochemistry* (the latest in this series (Kauranne *et al.* 1992) describes regolith exploration geochemistry). Very briefly, the sampler moves to a sampling site and, using a variety of methods, such as digging with a spade, using a portable drill (for overburden samples) or collecting a rock sample with a hammer, collects a geochemical sample which is given a unique sample number. This sample is then sent to the laboratory for analysis. The results are reported to the collector who uses the information, together with details collected in the field, to predict the occurrence of mineral deposits.

The most important item recorded in the field is the sample location but, typically, other relevant details of the site are recorded, particularly those that are likely to affect the geochemistry, such as depth for an overburden sample. Samples may differ in colour and texture and in some cases this can be important information used in the interpretation. A leached, white sample will be significantly different in chemical composition to the adjacent, red, iron-enriched horizon. Information is therefore collected about the sample as well as the location. BGS has been collecting this field information since 1970 on field cards that were designed to record the relevant sample information and location on a simple pre-printed form held in a small ring binder or Filofax (Fig. 1 shows the latest version for rock samples). Three forms were used in the original system, one each for soil, rock and drainage samples. These forms replaced the traditional field notebook, and information on the forms was intended to be stored or used in a computer.

Samples are normally sent to the laboratory in batches, which are prepared in the same way and analysed by the same methods. Nearly all geochemical laboratories record a batch number that is used to identify the group of samples through the analytical system and this batch number is a useful index to the samples. Chemical elements can be determined by a variety of methods and the results can be different depending on the method. These methods may also have different detection limits, bias or interferences. It is therefore important to record the method of analysis of each batch. This extremely brief introduction gives some indication of the data that need to be recorded in a geochemical database.

The flow of data in the MRP system can be shown in a simplified flow diagram that gives an overview, from sample collection in the field to final MRP report (Fig. 2).

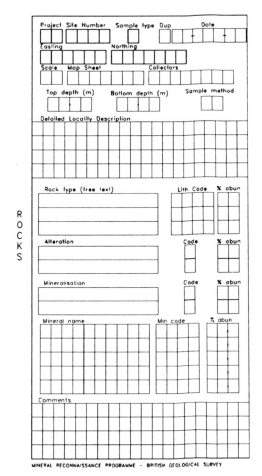

**Fig. 1.** Field data collection form for rock samples (1991 version).

## MRP database

The current MRP database was designed in 1986 by K. A. Holmes and J. S. Coats, and subsequently enhanced in 1989 by J. R. Harris with the addition of a user-friendly front end. The introduction of a relational database management system (ORACLE) in 1986 allowed data files, which has been previously kept in a flat-file database in an internally designed programming system G-EXEC (Gill & Jeffery 1973), to be managed in a modern software environment. The MRP model was based on a one-to-one correspondence between sample type and database table. Thus, there are tables for the four main types of samples collected: rocks, drill cores, soils and drainage samples (Fig.3). The last table contains details about several physical

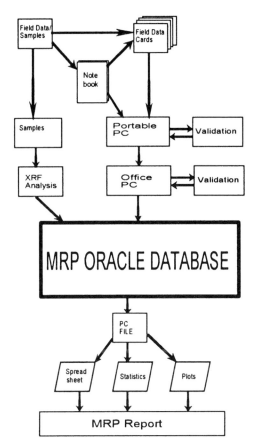

**Fig. 2.** Flow diagram for data and samples collected by the MRP.

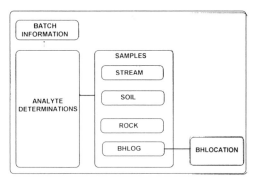

**Fig. 3.** Entity–relationship diagram for the MRP database model.

samples (steam sediments, pan concentrates and water samples) collected at the same site. Two other tables hold the borehole locations (BHLOCAT) and the analytical results (ALLEL).

The borehole location table (BHLOCAT) contains the locations of the boreholes and such information as the elevation of the collar and the date of drilling. The relationship to the bore sample table is maintained by the borehole's name. A better primary key would be the BGS registered boreholes number but at the time of database design this was not available in a database table.

The ALLEL table holds the analytical results in a de-normalized format with a single column for each chemical element. To allow for determinations of the same element by different methods, the primary key includes the method of analysis. The relationship between the sample tables and the analytical table is one-to-many.

Each sample can be analysed by many methods and a single field sample can have a duplicate analysis by the addition of a duplicate key. The relationship to the sample table is maintained via a foreign key which is optional to allow for the entry of analytical data without the corresponding field data and, a field sample does not have to be analysed.

Access to the tables is via user views which enforce security. Screen forms have been designed to allow user data entry and updating. These forms also maintain a full audit trail of all updates and deletions of data. Bulk loading of data is performed by FORTRAN programs which ask the user to describe the format of the input data file and name the columns of the table to be loaded. A menu-driven front end written in VAX DCL allows the non-SQL (Standard Query Language) user to navigate the forms and data entry programs. As the number of tables is relatively small, most SQL retrievals are simple to construct and, with training, most users of the database can construct their own queries. Example SQL retrievals are given in a manual issued to every user.

Several deficiencies have been identified in the present model and many of these are the result of the lack of full normalization of the data tables. For example, in the STREAM table there is a text field that lists the minerals identified in the heavy mineral concentrate. Because of the lack of normalization, this field can only be searched inefficiently as a variable length text field. Also, problems are encountered with trying to perform searches for 'GOLD' and, also, for synonyms such as 'Au' or 'AU'. This repeating group error is common in the field cards, where a number of items need to be recorded about the sample or site. Another example is catchment geology, which can be composed of several rock types, but is only recorded here as major and minor rock type. Another deficiency concerns the inability to locate and validate all the different types of samples within a certain area. The sample locations are split between four sample tables, SOIL, ROCK, STREAM and BHLOCAT, and searching can be laborious.

A particular difficulty is found in the ever-increasing width of the tables. In the existing database if a new element is determined on a batch of samples, for example a few samples are analysed for iridium, then a new column has to be added to the ALLEL table. In addition, as part of the design of the new model it was decided that each element determination also required a qualifier as many laboratories report determinations as greater than or less than a specific value. This would cause the width of the ALLEL table to double and therefore exceed the limit of 255 columns per table imposed by ORACLE. A similar problem exists with the STREAM table where the width increased with every new attribute that was added. This causes problems when sizing database tables as ORACLE does not guarantee that nulls will not be stored if a row is updated.

Another problem with the existing database is that the coded data on the field cards have evolved since the time of their introduction in 1970. As these codes were not identical it is impossible to carry out a retrieval without specifying the year of collection and also knowing the coding system employed in that year. Retrievals on data going back over a period of a few years can be very difficult. It was therefore decided to translate all past data into a new set of comprehensive field codes, and these '1990' codes would be adopted as the domains for the data analysis.

### New model

The aim of the new model was to describe accurately all BGS geochemical data for the UK landmass (excepting Northern Ireland) in order to facilitate the management of the data in a relational database. The new model was designed by the authors (Harris & Coats 1992) to meet this requirement and to overcome the deficiencies identified in the MRP system. During the redesign it was noted that the fundamental requirements and underlying structure of several other BGS geochemical datasets were very similar and they could be combined in one integrated BGS Geochemistry database. It was therefore decided to enlarge the aim of the new

database from just MRP data to all BGS UK landmass geochemical data.

A fundamental, but unstated, requirement of the first database design was that it should hold as many of the geochemical data as possible. The data model produced by the redesign is based on a superset of the data that exist and is, therefore, data-driven. A complete set of the field cards and the associated field handbooks, dating back to 1970, was collected. In all, ten sets of field cards were identified and with the extensive redundancy and complexity of the complete dataset it was decided to translate them all into a new set of field codes. Thus all the data would be stored in a single format that would make retrieval of a whole dataset more efficient and straightforward. A dictionary was compiled in the form of an ORACLE table containing every possible value of every field on the different cards and the values translated into the new set of '1990' codes (Harris et al. 1993). These new codes were adopted as the domains for the data analysis that followed and the attributes for each domain derived from the codes. These attributes were grouped into the initial relations similar to those in the original MRP database and then normalized to third normal form or higher.

A few of the general principles used in designing the new codes should be mentioned before describing the attributes. Because of the difficulties of validating free text and to enforce the recording of data in a format compatible across many sampling teams, coding of data was used wherever possible. The codes employed were those used in the existing forms or, where these could be shown to be defective, new ones were established. Hierarchical coding schemes were used wherever possible and preferably those published by other organizations or experts.

The entity–relationship (E–R) diagram for the logical model is presented in Fig. 5. The diagram

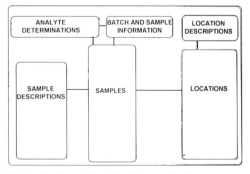

**Fig. 4.** Entity–relationship diagram for the geochemistry data model.

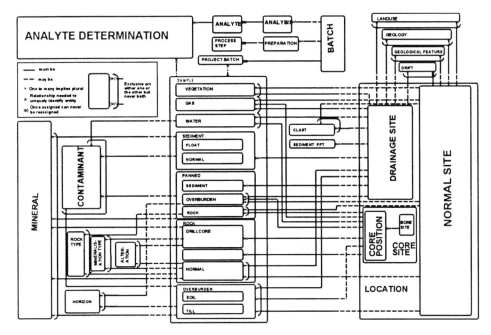

**Fig. 5.** Logical areas of the geochemistry data model.

can be divided into six logical area: locations, location descriptions, samples, sample descriptions, batch information and analyte determinations (Fig. 4). Each one of these areas will be considered in turn, giving the attributes recorded for each entity and the relationships. Detailed descriptions of the entities and domains are given in Harris & Coats (1992).

## Locations

Location is one of the key entities of a geochemical sample. Three subtypes of location are recognized: normal, drainage and core sites. An attempt was made to combine these into one entity but the information recorded at a drainage site, such as flow conditions or stream width, is very different to that at a rock sampling location. Borehole core locations are also different in that they are located by reference to the distance down the borehole, as well as the drill collar location. Normal and drainage location entities have *project* and *site number* as the primary key and the following attributes: National Grid easting and northing, grid accuracy, grid derivation, local grid east and north, elevation, top and bottom depth, map scale, map sheet, gamma count, solid angle and gamma environment, detailed locality, comments, collector and date collected. Additional

attributes recorded for normal sites are relief and profile drainage. Drainage sites have the extra attributes of drainage type, flow conditions, stream order, catchment area and weather.

Borehole sites are a special type of location that have National grid coordinates, a borehole name and number, drilled length, inclination, azimuth, drilling organization and name of core logger. The BGS standard reference number for the borehole is also added so that the table can be linked to the main BGS borehole database. The table linked to this location entity is core position, which has the project code and site number for the sampling site within the drilled length. This forms an alternate key to the table along with the primary key of borehole name, borehole number, top and bottom depth.

## Location descriptions

Linked to the normal and drainage site locations are tables describing the surrounding area for land use, geology, geological features (such as faults or mineral veins) and drift. Land use uses a domain of 102 hierarchical codes derived from the 1966 Land Utilisation Survey and includes both natural and industrial uses. The latter are clearly important in urban areas and in identifying possible contamination. The geology table uses existing BGS hierarchical codes for lithology (Harrison & Sabine 1970) and the attributes, stratigraphic age and relative abundance. The drift table uses a domain derived by classifying the commonly used units appearing on the margins of geological maps. It is expected that this area of the database will be revised when other subject areas of the BGS data architecture are defined in detail.

Linked to the drainage site are tables containing information on clast composition and abundance, and sediment precipitate, both of which are important in identifying possible sources of geochemical anomalies. Originally, the database model linked contamination to the drainage location entity, but it was felt that contamination should be linked to actual contamination of the sample, rather than to potential contamination at the site. Clast composition can also be linked to a normal site to allow for the recording of clasts observed in overburden profiles.

## Samples

Several sample types can be collected from one location and the relation to the location entity must be enforced. In other words, every sample must have a location but not every location has to be sampled. Also at a drainage site, samples of stream water, stream sediment and heavy mineral concentrate can be collected from the same location. The subtypes of sample are identified by the sample type code and the standard identification has been inherited from the original 1970 definition employed on the early fields cards (S = soil, C = stream sediment, P = panned concentrate, T = till, D = drill, etc.). This has some deficiencies but the difficulty of recoding many thousands of samples made this impossible to change. The sample inherits the project code and site number from the location table and, when combined with the sample type and duplicate codes, forms the primary key to the table.

## Sample descriptions

Entities linked to the sample are contaminant as described above, horizon, rock type (or lithology), mineralization, alteration and mineral. The BGS standard mineral codes are used for the mineral domain (Harrison & Sabine 1970) and these can be linked to many of the sample types. Using this technique the occurrence or content of all minerals in many different types of sample

can be managed and validated in one table and not scattered as attributes through many tables. The entity 'Horizon' uses a domain derived from the Soil Survey (Hodgson 1976) to describe the horizon that is sampled and has the attributes of Munsell colour, soil texture and soil type.

## Batch and sample information

Batches of samples are prepared and analysed and this subject area of the database forms a metadata index to the sample analyses. Samples are grouped into project batches which have the same project code and sample type. The maximum and minimum site numbers, and the total number of samples, are recorded. This allows summary information on the range of samples included in one batch to be retrieved. An entry in the batch table may be made up of many project batches, thus allowing samples from several project areas to be combined. The batch entity has a primary key, composed of the batch identity and the laboratory, and attributes of date registered, geographical area and locality, which is validated against the Ordnance Survey gazetteer. Using this field, searches such as 'the batch of rock samples collected in Aberfeldy in 1980' can be completed. The analysis table holds information on the date and method of analysis of the batch, and the elements or species determined by that method are held in the analyte table, which contains quality control data such as the limits of detection. Other quality control data, such as precision, can be linked to this table by the analyst.

Originally there was the intention to model the complete laboratory analysis and preparation system, but this is a complex logical area and does not offer many benefits in added information value to the aims of the database. A Laboratory Information Management System (LIMS), which interfaces the analytical instruments and communicates with the Geochemistry database, is more appropriate for this task and is currently under evaluation in BGS.

## Analyte determinations

The analyte determination table is the largest in the database and it is estimated to contain six million element determinations when fully populated. Because each row contains only sample number, method of analysis, laboratory, batch identity, analyte and abundance, the table is fully normalized and contains no null abundance values.

Retrieving data in tabulated form from this kind of structure was examined in detail and can best be achieved using SQL by decoding on the method of analysis and the analyte and then grouping by the unique sample identifier and averaging the values to obtain one row per sample. The average has no effect as it is the average of a series of nulls and one value, and is used merely to obtain the desired format (Harris & Wild 1992). Determinations are all stored in the same units (ppm by weight) to maintain uniformity and to avoid confusion. Parts per million was chosen as the best compromise between percentages, which would involve many zeros, and parts per billion, which would involve very large numbers. The majority of elements have Clarke values in the 1–10 000 ppm range (Mason & Moore 1982). The external schema (the view of the database seen by the user) may have per cent oxide for the major elements but retain the concentrations in the logical schema in parts per million.

## Referential integrity

Site and sample number are of crucial importance to the data model because they are inherited as the primary key of most of the tables. The first work reported after the data design (Harris & Coats 1992) was concerned with an analysis of the sample numbering systems that had been used by BGS (Harris et al. 1992a). These systems have shown remarkable consistency over the long life of the Survey and, apart from the temptation of every new recruit to BGS to design their own numbering system, have been very successful in mineralogy and petrology, geochemistry and geology. Separate, but very similar, systems have been established in each of these areas and, by using a single code to identify the numbering system, all the samples collected by BGS can be uniquely identified. Rules have been developed to police the numbering system and to renumber samples that break them. Because the database captures much of the analytical data before it reaches the geologist, it can enforce these rules and prevent duplicate sample numbers being entered. A full record can also be kept of any subsequent renumbering.

Batch information is also critical to the database design as it provides a management and quality assurance tool and, also, a metadata index to all the analytical data produced by BGS. Sample analyses without batch information may suffer from a variety of problems, such as missing analytical methods, unknown laboratories, variable precision or detection limits and even post-analysis 'normalization'.

## Discussion and conclusions

The logical model presented in this paper is capable of describing the geochemical data held by BGS for the UK land surface and thus meets its primary aim. With the addition of latitude and longitude, and further fields to describe sea-bed sediments, it should also be able to hold data pertaining to offshore data. Geochemical data from elsewhere in the world can be largely encompassed by the model (Jones & Coats 1993) but difficulties arise because of the lack of control over the site and sample numbering system. A relational database cannot allow duplicate primary keys in its tables and rules must be devised to add an extra attribute for the country of origin.

A difficulty with the new data model is that it does not include bibliographic references. This was not included in the logical data design and did not feature as a strong requirement of the users interviewed before proceeding to the physical design (Harris et al. 1992b). The numbering system allows the identification of the main project under which the sample was collected, but does not reference the report in which the result was reported. As a single reference may report many analyses and an analysis may be reported in several references, this is a many-to-many relation. The number of sample analyses to be held by the database is predicted to exceed six million, and the linking table to hold this relation would be very large. Other smaller tables such as the project batch are possible candidates but are logically difficult to join to references. As the requirement for a link to bibliographic references is only small and chiefly confined to lithogeochemical samples that form only a small minority in the database, the relation has not been implemented.

The geochemistry data model described in this paper is broadly compatible with the BGS data architecture (Logica 1991) but there are differences. In the geochemistry data model the site or location number is inherited by the sample (because geochemists usually collect at least one sample from every site they visit). In the BGS data architecture, site and sample numbers are different (because a geologist mapping an area visits many locations but collects only a few samples). This difference is not yet a problem because the databases are not closely linked but may be a difficulty in the future.

This paper is presented partly as a case history of seven years' experience in operating a relational database to hold geochemical data and, also, to present a new data model for geochemistry. Co-operation in other areas of geology, such as oil exploration, has led to the development of common standards for relational data models (Petrotechnical Open Software Corporation 1993) and this paper is a contribution to a future debate on standards for data models in geochemistry. At present, where large datasets from many different exploration companies have to be combined, only the simplest form of data structure, that of a flat-file data table of analytical results, is possible. The evolution of common standards for relational data models will allow much more sophisticated data integration and analysis.

The authors would like to thank their colleagues in BGS for their co-operation in this work and the Sandwich Students from the local Universities who have helped in the implementation of the database. The authors publish by permission of the Director of the British Geological Survey (Natural Environment Research Council).

# References

COATS J. S., SMITH, C. G., FORTEY, N. J., GALLAGHER, M. J., MAY, F. & McCOURT, W. J. 1980. Stratabound barium-zinc mineralisation in Dalradian schist near Aberfeldy, Scotland. *Transactions Institution of Mining and Metallurgy (Section B Applied Earth Science)*, **89**, B109–122.

COLMAN, T. B. 1990. *Exploration for Metalliferous and Related Minerals in Britain: A Guide*. British Geological Survey, Nottingham.

GILL, E. M. & JEFFERY, K. G. 1973. *A Generalised Fortran System for Data Handling*. Institute of Geological Sciences, Computer Unit, Bulletin **73/3**.

HARRIS, J. R. & COATS, J. S. 1992. *Geochemistry Database: Data Analysis and Proposed Design*. British Geological Survey Technical Report **WF/92/5** (BGS Mineral Reconnaissance Programme Report 125).

——, GLAVES, H. & COATS, J. S. 1992a. *Geochemistry Database Report 2: A Proposed Integrated Sample Numbering System*. British Geological Survey Technical Report **WP/92/5R**.

——, COATS, J. S., WILD, S. B., KHAN, A. H. & STEPHENSON, P. 1992b. *Geochemistry Database Report 4: The User Requirement*. British Geological Survey Technical Report **WP/92/12R**.

——, NICHOLSON, C. J. & COATS, J. S. 1993. *Geochemistry Database Report 6: Standardisation of Geochemical Field Cards 1970–1992*. British Geological Survey Technical Report **WP/93/20R**.

—— & WILD, S. B. 1992. *Using the Oracle V.6 Explain Plan Utility*. NERC Computing, No. 54, 22–26.

HARRISON, R. K. & SABINE, P. A. (eds) 1970. *A Petrological–Mineralogical Code for Computer Use*. Institute of Geological Sciences Report **70/6**.

HODGSON, J. M. (ed.) 1976. *Soil Survey Field Handbook*. Soil Survey Technical Monograph No. 5. Soil Survey, Harpenden, Herts.

JONES, R. C. & COATS, J. S. 1993. *Sources of Geochemical Data Collected by the International Division since 1965*. British Geological Survey Technical Report **WC/93/17**.

KAURANNE, L. K., SALMINEN, R. & ERIKSSON, K. (eds) 1992. *Regolith Exploration Geochemistry in Arctic and Temperate Terrains. Handbook of Exploration Geochemistry Volume 5*. Elsevier, Amsterdam.

LEAKE, R. C., CAMERON, D. G., BLAND, D. J., STYLES, M. T. & ROLLIN, K. E. 1992. *Exploration for Gold in the South Hams District of Devon*. British Geological Survey Technical Report **WF/92/2** (BGS Mineral Reconnaissance Programme Report 121).

LOGICA COMMUNICATIONS LTD. 1991. *British Geological Survey Data Architecture Study*. British Geological Survey Database Documentation Series Report.

MASON, B. & MOORE, C. B. 1982. *Principles of Geochemistry*. 4th Edition, Wiley, New York.

PETROCHEMICAL OPEN SOFTWARE CORPORATION 1993. *Epicentre Data Model Version 1.0*. Prentice-Hall, New Jersey.

# The nature of data on a geological map

J. R. A. GILES & K. A. BAIN

*British Geological Survey, Kingsley Dunham Centre, Keyworth, Nottingham NG12 5GG, UK*

**Abstract:** The primary purpose of the traditional geological map is to convey the map maker's understanding of the geology of an area to the user of the map. A second, and less well understood role is as the *de facto* repository for much of the recorded geological data of an area. With the advent of digital geological map production, geographic information systems and digital geological databases, these traditional roles are changing. The data from which a map is constructed can reside in geological databases and the maps produced from these databases can concentrate on conveying information to the map user.

The design and construction of the digital geological databases used to store the information collected by geologists requires a detailed understanding of the elements of a geological map. This paper attempts to describe those elements and discusses some of the problems caused by the conventions used on geological maps.

The British Geological Survey (BGS) has been investigating digital map production for over 20 years (Bickermore & Kelk 1972; Rhind 1973; Loudon *et al.* 1980, 1991; Loudon & Mennim 1987). Commonly, these investigations were conducted in parallel with on-going regional mapping projects. Each investigation asked the same basic questions: could commercially available software and hardware produce geological maps of the type, quality and scale required by BGS and, if so, would it be cost-effective to use them? At the end of each investigation the technological, and to some extent methodological problems, that had prevented a successful outcome were identified. New projects were started when vendors could supply solutions to the problems that had been encountered in previous projects, but only with the completion of the Wrexham Development Project (part funded by the Department of the Environment) in 1990 (Loudon *et al.* 1991) was it decided that the time was right for BGS to develop a system for routine production of digital maps. The Digital Map Production Implementation (DMPI) project (Nickless & Jackson 1994) was therefore set up and started in April 1991 with the aims of developing:

- a digital database of spatial and non-spatial information;
- a system for the rapid production of high-quality 1:10 000-scale geological maps directly from a digital environment;
- the capacity to revise maps by manipulation of the digital database;
- a consistent format for map presentation throughout BGS, for the standard products;
- the capacity to provide customers with geological map data in digital form;
- the capacity to produce non-standard, customer-specified output.

Maps produced digitally provide numerous advantages both to the customer and to BGS. At the simplest level, colour electrostatic plots of 1:10 000-scale geological maps of good cartographic quality will replace the geologist's hand-drawn map, known within BGS as a 'standard'. Monochrome dyeline copies of such standards have been the only publicly available map at this scale since routine lithographic printing ceased in the 1960s. Hand colouring was commonly essential to make them easier to understand. The digital product is far more attractive and flexible. Maps can be produced in colour or monochrome, at a variety of scales covering any area including customer-specified site-centred maps.

The system is made even more flexible when the digital information is attributed geoscientifically. During the attribution process, a geographic information system (GIS) is used to link individual objects depicted on the geological map to a relational database. Such geographic objects could, for instance, be a length of geological boundary or a structural observation. Once these geoscientifically attributed data are captured within the system it is possible to select from them single- or multi-theme sub-sets of the information to make specialized geological maps to meet specific customer needs.

## System description

The digital map production system comprises two principal pieces of software mounted on

*From* Giles, J. R. A. (ed.) 1995, *Geological Data Management,*
Geological Society Special Publication No 97, pp. 33–40.

appropriate platforms: the GIS and the relational database. These are supplied by different vendors who do not specifically target their product at the geological market sector. It was critical to the success of the project that these two components worked together. Each can store the various items of information depicted on a geological map, but they store it in different ways. Geological information represented on a map has two types of information related to it. There is the information about the position and areal extent of the object on the map, such as a length of a geological boundary or the area over which a given rock type is exposed. This type of information is best held in a GIS. The second type of information describes and qualifies the geological objects depicted on a map; for example the type of geological boundary, whether it is intrusive or unconformable, or the definition of the rock unit. This type of information is ideally suited to a relational database.

A logical model of the data in the system was built to aid the understanding of what is shown on a geological map and to ensure that the components of the system would work together. The identification and understanding of the different types of data on geological maps was essential if the system were to fulfil its primary function, production of geological maps from a digital database. It was also considered important to have an overall logical model of the data to provide a degree of vendor independence. It is a common error to spend large sums of money on equipment and software only to find that control of the business objectives has been lost to the principal supplier of the hardware and software. To avoid this, the logical model was developed and will be maintained. The model allows system implementation on any one of a number of hardware platforms with various software configurations.

The methodology used on the project was adapted from the Learmonth & Burchett Management Systems Structured Design Methodology. The work was aided by use of the Computer Aided Software Engineering (CASE) tool SYSTEMS ENGINEER. The following steps were used:

- examine source documents to develop provisional entity relationship diagram;
- define entities;
- allocate attributes to entities;
- define attributes;
- verify the entity relationship diagram by normalization;
- have the model reviewed by appropriate staff.

Data modelling was one of the most powerful techniques used. Its objective is to produce a representation of the data in the system, in this case the data presented on a geological map. The model records two main features of a system. First of these are the system entities, the objects about which the system needs to record information such as a geological boundary. These are represented by rectangular boxes (Fig. 1). There are several types of entity but the most common are master entities and detail entities. The former, such as the position at which an observation was made, can exist independently of other entities. The latter, such as the record of an observation made at a given position, describe some property of another entity.

Secondly, the model records information about the relationships between entities. These are represented by lines between the entities which have additional symbols attached. Three symbols are used: a circle, a bar and a crowsfoot. These indicate the cardinality of the relationship. The outer pair of symbols, nearest to each entity, indicate the maximum cardinality. A bar indicates a maximum cardinality of one whilst a crowsfoot indicates a maximum cardinality of many. The inner pair indicate the minimum cardinality. A bar indicates a minimum cardinality of one and a circle indicates a minimum cardinality of zero. So in Fig.1, a given position may be described by zero or many observations, whilst an observation must describe one, and only one, position.

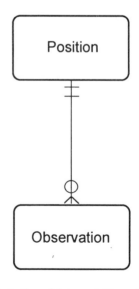

**Fig. 1.** Illustration of data modelling notation.

## Data on a geological map

The geological map is a tool, the primary function of which is to convey a geologist's interpretation of the geology of an area to the map user. It also acts as a storage medium for some of the factual geological data for a given area.

The map face has two classes of information displayed upon it. The first is the data that are essential to the explanation of geology, such as boundaries, areas and structural observations. The second class consists of non-essential information which informs the map user about the range and types of data sources available to the compiler of the map. The category includes items like borehole, shaft and adit locations, fossil or petrographical slide localities, or information from underground mining sources. Both classes of information can only be interpreted by reference to the map marginalia. The following discussion does not distinguish between these two classes of data.

The geological map is a highly stylized document that uses a variety of explicit symbols and contextual information to fulfill its tasks. In many ways it is akin to a novel. The author uses language, enriched with metaphors and similes, to convey the description of a landscape to the reader. Readers do not see the landscape physically, but understand it and can picture it in their mind. In a two-stage process the author describes the landscape in words and the reader interprets what was written. This process relies upon the author and the reader being literate in a common language.

Geologists are taught to prepare and interpret geological maps from the earliest days of their careers (Boulter 1989; Maltman 1990; Powell 1992). An essential part of their training is familiarization with the conventions; so complete is the 'indoctrination' that they more or less forget that conventions exist.

The conventions used on a geological map, like those in any language, have evolved to suit the way in which the human brain receives information from the world around it and converts it into concepts that are stored and manipulated. The conventions regulate the message passed from the author to the user, in an attempt to minimize misunderstanding. However, information in a form suitable for human use may be unsuitable for a computer. Examples of this conflict are discussed below.

### Explicit geological data

The simplest form of data on the map is unambiguous, explicit, commonly positional information, associated with a numerical value or character string. This type indicates the presence of a feature such as the location of a borehole, photograph or rock specimen. It is usually accompanied by a unique identifier such as the borehole number or the collection number. This is represented in a data model by a one-to-many relationship to the position at which the data were collected. The only complication is when an observation has multiple identifiers, such as boreholes registered in several collections. In this case a further one-to-many relationship may be necessary (Fig. 2).

### Less explicit geological data

Not all data on the map are as explicit as they first appear. Structural geology symbols provide positional information similar, in many respects,

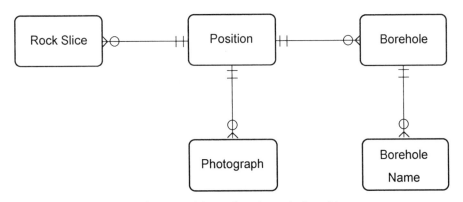

**Fig. 2.** The relationship between various sets of data collected at a single position.

to that provided by the borehole symbol. For example, a dip arrow on a map marks the position at which an observation was made. It also depicts the direction of dip but does not quantify the azimuth. The amount of dip may be given by a number associated with the head of the dip arrow. This presentation of the information makes it easy for the map user to visualize the information present even though they have to estimate the dip azimuth. The dip azimuth value was recorded in the field and used to plot the symbol, but this information has been devalued in the process of making it more available to the user. To make best use of the information held within data the human mind needs the data to be presented in a form more immediately intelligible than that which is suitable for storage in a computer.

The dip arrows and dip values presented in Fig. 3 are readily interpreted by a geologist as illustrating a northward-plunging synform. The list of original observations in Table 1 is much more likely to be how the computer stores the data for maximum flexibility.

Different types of information may be combined within a single symbol. For example, the dip and strike symbol may be combined with the symbol that indicates overturned strata. This symbol describes a combination of two different properties. The dip is a measurement made at a point, but the way up of the rocks is an interpretation of the wider structure. Thus a single symbol on the map associates a 3D vector property of the bedding (dip) with the age relationships of the rocks immediately above and below (the way up). Such arbitrary combinations of information must be disentangled and made explicit before entry into a digital database.

## Contextual geological data

Some elements of the data on the geological map are contextual; their meanings are dependent upon the surroundings in which they are displayed.

The intersection of a body of rock with a surface (such as a rockhead) is depicted by a symbolized polygon bounded by symbolized lines. Various symbols are used to distinguish the different types of geological boundaries, e.g. a solid/broken bold line for a fault. The most commonly met symbol represents 'Geological boundary, Solid'. It is a catch-all symbol used to classify all the types of solid geological boundary not otherwise identified by different symbols in the map key.

Further information about the exact nature of the geological boundary can be elucidated from a study of the map and the generalized vertical section. If the geological units on either side of

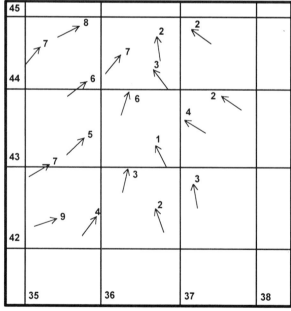

**Fig. 3.** Methods of presenting information appropriate for humans.

**Table 1.** *Methods of presenting information appropriate for computers*

| Grid reference | Azimuth (degrees) | Dip (degrees) |
|---|---|---|
| 3664 2449 | 301 | 2 |
| 3641 4396 | 022 | 6 |
| 3728 4481 | 311 | 2 |
| 3762 4391 | 308 | 2 |
| 3528 4461 | 028 | 7 |
| 3579 4334 | 019 | 5 |
| 3547 4231 | 081 | 9 |
| 3665 4331 | 333 | 1 |
| 3672 4477 | 348 | 2 |
| 3632 4448 | 032 | 7 |
| 3573 4487 | 054 | 8 |
| 3667 4432 | 331 | 3 |
| 3595 4237 | 029 | 4 |
| 3709 4364 | 310 | 4 |
| 3582 4412 | 023 | 6 |
| 3535 4305 | 074 | 7 |
| 3637 4299 | 009 | 3 |
| 3719 4285 | 355 | 3 |

the line have been deposited at significantly different times, the line may represent a disconformity. If two units have different dips, the line between represents an angular unconformity. If the dips are the same and the units were consecutive, the boundary is conformable.

The exact meaning of the line is not indicated explicitly by symbols on the map, but is readily discernible by an experienced map user either intuitively from their knowledge of geology or by reference to information in the margins of the map. For example, a geological boundary on the map that represents a disconformity can only be interpreted on the basis of regional knowledge or by reference to the vertical section. For digital storage the contextual information needs to be made explicit.

Another common practice on geological maps is to identify a polygon as a specific lithostratigraphical unit by means of a symbol. Within each such polygon other polygons may be identified as specific lithological variations of the lithostratigraphical unit. For example, the Scunthorpe Formation comprises calcareous and silty grey mudstones interbedded with feature-forming limestones. On a geological map the limestones are shown within the Scunthorpe Formation polygons but are identified only by the symbol 'L'. They are not explicitly labelled as Scunthorpe Formation, but this is inferred by their context. This practice is only confusing on the map face if the lithological unit is at the base or top of the lithostratigraphical unit to which it belongs, but

the interrelationships will be explained in the generalized vertical section in the map margin.

## Multivalued data

An example of multivalued data, an extension of contextual data, is presented by geological lines. A fault which separates two lithostratigraphical units is the common boundary to both units. From whichever side of the line you approach the shared boundary it is always a fault. Even though one side of the line is the upthrow side and the other is the downthrow side, the line itself has a single value. A line which is multivalued is an angular unconformity, such as the boundary between the Middle Coal Measures Formation and the Permian Basal Sand. The one line is a single symbol, but means two things depending upon the side from which it is approached. In the digital environment these alternatives must be made explicit if maximum use is to be made of the available information.

## Sidedness

Something that the human brain can cope with far better than a computer implementation is a property which belongs to one side or the other of a geological feature. The most common example of this is the hatchmark (tick) on a fault, indicating the downthrow side. This mark is meaningful only when taken in the context of the surrounding information. These types of data can be handled by software which deals with lines as 'directed arcs', i.e. having different properties going from A to B as opposed to from B to A, but the consideration is one of implementation rather than analysis.

## Structured data

On geological maps lithostratigraphical and lithological symbols carry much information about the polygons they label. The primary function of the symbol is to form a link between the object on the map face and the explanation of what that object represents, in the map key. In some cases the map key carries only a minimal explanation of the map symbol. In the case of the Scunthorpe Formation a typical entry in a map key would be 'grey mudstone with minor limestone beds'. Much more information may be available off the map sheet if a formal definition in terms of the North American Stratigraphic Code (1983), has been published. The definition of the Scunthorpe Formation was published by Brandon *et al.* (1990) and its hierarchical relationships explained. The Scunthorpe Formation is a component of the Lias Group and comprises five members, including the Barnstone Member. This sort of structured data may be available in the map key, but it must be stored within the database to obtain maximum benefit from digitization. Database searches on Lias Group strata will retrieve all Scunthorpe Formation and Barnstone Member polygons as well as those identified as Lias Group. This can be achieved by using a data model structure which has multiple one-to-many relationships between entities (Fig. 4). Such a representation is known as a 'bill of materials' structure because it is frequently used to model the complex hierarchial relationship between components and finished products in a manufacturing environment (Barker 1990).

## Pseudo-3D geological data

To geologists, a geological map provides a mental image of the three-dimensional disposition of the depicted rock units even though the geological map may contain very little specific 3D information.

The full extent of the intuitive and interpretive processes that geologists perform when they examine a map only becomes apparent when the data presented on a geological map are used as a source for 3D modelling. Unlike a geologist, a computer needs fully coordinated 3D data to produce a 3D model. Some data may be produced by extrapolation from the available data points, say the intersection of geological boundaries with contours, but this will generally be insufficient to make a reliable model itself, whether it is made up of stacked surfaces or voxels.

The items of 3D data presented on a geological map consist of selected information projected from significant geological surfaces onto the plane of the map. A significant geological surface is one on which the geology is interpreted or on which observations are recorded. For example, the solid geology is interpreted on the rockhead surface. This may be coincident with the topographical surface or be buried beneath a thickness of unconsolidated sediments. The interpreted geology of the rockhead surface is projected onto the plane of the map. Other surfaces are treated in the same way. The upper surface of the unconsolidated deposits is more commonly coincident with the topographical surface; but in many urban areas

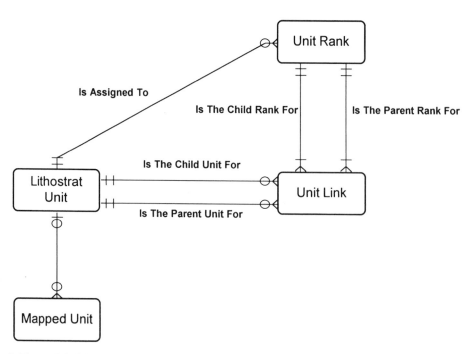

**Fig. 4.** The model of the relationship between lithostratigraphical units for a given polygon on a geolocial map.

it may be buried beneath man-made deposits. It too is projected onto the plane of the map. Other items of 3D information are also projected from the surface on which they were recorded onto the plane of the map, e.g. seam contours recorded in coal mines. The net effect is to kaleidoscope the information from many different surfaces onto the single plane of the map.

The most common of these significant surfaces are:

- topographic surface;
- drift surface;
- rockhead surface;
- unconformities such as the sub-Permian surface;
- coal seams;
- metalliferous mineral workings.

Lithostratigraphical units on the map are shown as polygons defining their areal extent where they intersect a geologically significant surface. When the unit has no drift cover, the rockhead and the topographic surface are coincident. When the unit is covered by thick drift deposits, such as the river terraces and alluvium of a river valley, the drift and rockhead surfaces may be many tens of metres apart. A polygon depicting the distribution of a drift deposit, such as alluvium, is shown on the drift surface and projected onto the plane of the map.

Boundaries to lithostratigraphical units are shown by lines on the appropriate geologically significant surfaces. Boundaries of solid units are depicted on rockhead surfaces projected onto the plane of the map. Likewise the boundaries of unconsolidated sediments are depicted on the drift surface and in turn projected onto the plane of the geological map. (Water bodies, such as lakes, are always excluded from the solid and drift surfaces; these surfaces form the base of the water body.) Where the boundary is a fault it may be depicted on several geologically significant surfaces. It will be shown on the rockhead surface as a boundary of a lithostratigraphical unit and as a line on underground surfaces such as the floor of one or more coal seams.

By convention a geological map is built up from the projection of the geology of significant surfaces onto the plane of the map. As a direct result information can be totally lost if the rock body does not intersect one of these significant surfaces.

For example, a rockhead channel filled with sand and gravel but buried by till will not be depicted on a geological map at all. Such sand and gravel may be a significant resource. The till that overlies the sand and gravel is shown on the map because it intersects the drift surface, and the solid rock beneath it is shown because it intersects the rockhead surface. The 3D distribution of the hidden deposit may be well understood as a result of borehole or geophysical surveys, but because it does not intersect one of the geologically significant surfaces it is not depicted on the map.

The interrelationship between the 3D distribution of a rock body and the geologically significant surfaces used to create the map determines whether a given unit is shown on a map. For special purpose maps, such as resource maps, non-standard significant surfaces may be more meaningful than those used by convention. In the above example the sub-till surface would have depicted the distribution of the sand and gravel resource.

## Conclusion

The geological map includes selective information presented in a form that can be assimilated by an experienced user into a 3D conceptual model of the geology of a region. Its primary purpose is to present information to a human user. Its use requires intuition, past experience and intelligent assessment. Most of the data on a geological map cannot be used directly by the computer for the same purpose. For the database, ways have to be developed of making all the data explicit so that they can be stored digitally.

The development of a digital map production system that is built around a geological database requires an analysis of geological maps, the data on them and the data from which they are constructed. The most difficult task is to analyse and classify map data so that they can be used in a digital system.

The primary task of the DMPI project was to create a system that could produce a geological map, but the database that has been created should lend itself to other scientific purposes, including 3D modelling.

Our thanks to colleagues in the Data and Digital Systems Group and the Information System Group of the BGS for their constructive criticism and to all the staff of the Digital Map Production Implementation Project. S. Henley and an anonymous referee are thanked for their helpful and encouraging comments. The work is published with the approval of the Director, British Geological Survey (NERC).

## References

BARKER, R. 1990. *Entity Relationship Modelling*. Addison-Wesley Publishing Company, Wokingham.

BICKERMORE, D. P. & KELK, B. 1972. Production of a multi-coloured geological map by automated means. *Proceedings of Twenty-Fourth International Geological Congress*, **16**, 121–127.

BOULTER, C. A. 1989. *Four Dimensional Analysis of Geological Maps*. Wiley, Chichester.

BRANDON, A., SUMBLER, M. G. & IVIMEY-COOK, H. C. 1990. A revised lithostratigraphy for the Lower and Middle Lias (Lower Jurassic) east of Nottingham. *Proceedings of the Yorkshire Geological Society*, **48**, 121–141.

LOUDON, T. V., CLIFTON, A. W., GILES, J. R. A., HOLMES, K. A., LAXTON, J. L. & MENNIM, K. C. 1991. *Applied Geological Mapping in the Wrexham Area—Computing Techniques*. British Geological Survey Technical Report **WO/91/1**.

—— & Mennim, K. C. 1987. *Mapping Techniques Using Computer Storage and Presentation for Applied Geological Mapping of the Southampton Area*. Research Report of the British Geological Survey No. **ICSO/87/3**.

——, WHEELER, J. F. & ANDREW, K. P. 1980. A computer system for handling digitized line data from geological maps. *Computers and Geoscience*, **6**, 299–308.

MALTMAN, A. 1990. *Geological Maps: An Introduction*. Open University Press, Milton Keynes.

NICKLESS, E. F. P & JACKSON, I. 1994. Digital geological map production in the United Kingdom—more than just a cartographic exercise. *Episodes*, **17**, 51–56.

NORTH AMERICAN COMMISSION ON STRATIGRAPHIC NOMENCLATURE 1983. North American Stratigraphic Code. *AAPG Bulletin*, **67**, 841–875.

POWELL, D. 1992. *Interpretation of Geological Structures through Maps: An Introductory Practical Manual*. Longman Scientific & Technical, Harlow.

RHIND, D. W. 1973. *Computer Mapping of Drift Lithologies from Borehole Records*. Report of the Institute of Geological Sciences, **73/6**. HMSO, London.

# Improving the value of geological data: a standardized data model for industry

MICHAEL R. SAUNDERS[1], JOHN A. SHIELDS[2] & MICHAEL R. TAYLOR[3]

[1] Baker Hughes INTEQ, Shirley Avenue, Windsor, Berkshire SL4 5LF, U.K.
[2] Baker Hughes INTEQ, 7000 Hollister, Houston, Texas 40265, USA
[3] Baker Hughes INTEQ, 17015 Aldine Westfield Road, Houston, Texas 73118, USA

**Abstract:** The preservation of observed geological data in a standardized and digitally accessible form presents a challenge. Geological data objects range in scale from the macroscopic, e.g. outcrop and core, to the microscopic, e.g. drill-cuttings. Each change in scale has a direct effect on apparent heterogeneity, which may result in contradictory data, and the providers of the observations may have widely differing degrees of experience. A system has been developed for the definitive description and preservation of features in core and drill-cuttings using workstations. It provides the opportunity to extract definitive individual attributes from geological descriptions with a minimum of extra effort. These digitally stored data are represented graphically, textually and symbolically to construct geological logs. Geological data may be presented in combination with drilling and mud gas data, Measurement While Drilling (MWD) and wireline logging data, and off-site analyses, to produce composite formation evaluation logs, as well as being exported in customized file formats, thereby enhancing the value of all data through integrated information management.

The data collected from drilled samples fall into two general categories:

- cursory, and/or detailed, petrological descriptions of cuttings or core made while drilling, to make wellsite logs and reports; the detail and validity of which depend upon rate of sample delivery and the describer's skill;
- detailed post-drilling analyses from specialist laboratories.

Most wellsite derived descriptions are first recorded in detail on worksheets or in note-books, elements from which are often transferred to computers. The most significant advances in recent years have been related to the process by which hand-drawn logs have been superseded by computer-drafted logs. As the face of the computers has become friendlier, data recording interfaces have been developed to facilitate ease of access, usually based on effective but restricted palettes or electronic notebooks (Spain et al. 1991). Unfortunately the data are intrinsically hard to access by third parties and are reduced to traditional graphic logs for most purposes.

In these applications of geological information technology only those elements which must be available for plotting purposes are stored digitally. The record with the highest information density, the written description, is rendered as a text field which is only verifiable in terms of spelling, ordering and, correctness, by eye. The possibility that errors exist in the text make it unsuitable for automated data extraction or translation; in this way the computer-drawn log is not a significant advance in data availability over hand-drawn logs.

Consequently, the wellsite description of the freshest geological samples has generally been either ignored or rapidly superseded by later laboratory analyses. Its most lasting contribution beyond the drilling phase is usually as highly condensed annotation on composite logs. Rarely has the wellsite geological description been easily available to petrophysicists and reservoir engineers engaged in post-well processing.

As a result of extensive experience in providing these kind of data worldwide, the limitations restricting the long-term usefulness of the data were apparent to us. We felt that improvements we had made to data quality were being wasted because of the restricted access and subsequently reduced useful life-span.

The opportunity to address these issues arose from the need to develop a paperless worksheet for describing and logging slim-hole core to millimetre resolution, at rates of up to 150 m/day. The worksheet was part of a core logging system developed jointly with BP and STATOIL (Jantzen et al. 1993). Concurrently, our UNIX-based information management systems were being upgraded to take advantage of improvements in database technology (Morley 1990);

From Giles, J. R. A. (ed.) 1995, *Geological Data Management*,
Geological Society Special Publication No 97, pp. 41–53.

and so any model developed for core could be used for other samples.

An extensive geological data model, a prototype worksheet interface, and basic core log formats were developed and implemented during the field trials of the core logging system. The comments of the wellsite geologists were used to drive the system forward through its pre-commercialization stage. Upon completion of the trials the complete geological description system was further developed for use with drill cuttings, for use by both wellsite geologists and engineers.

## Geological data

Geological data are very variable, ranging from discrete objective data, like mineral names (e.g. sulphur) through descriptive continua, like colour, to more subjective interpretations such as hardness. In most examples of measurable continua, subdivisions have been superimposed on the data range and named. Some schemes are almost universally accepted, like Wentworth grain size (Wentworth 1922); other schemes have local adherents, e.g. Pettijohn versus Folk classifications of clastic rocks (Pettijohn 1954; Folk 1954). If geological data are to be useful then the precise meaning of any description must be understood by both describers and users.

### Data scale

The data that are of geological rather than strictly engineering significance fall onto a number of scales: there are the basin-scale features derived from geophysical techniques; the detailed mesoscopic studies performed on cores and outcrops (e.g. sedimentology, geochemistry, palaeontology and petrophysics); and there are the meso- and microscopic determinations performed at the wellsite by mudloggers and wellsite geologists. Most of the data modelling effort to date within the oil industry has focussed on the bigger picture. We turned our attention to refining the data model from the bottom up, with the ultimate goal of making wellsite-gathered descriptions at least as useful and long-lived as petrophysical data.

### Data sources

The information encompassed by the data model was to be derived from visual observation and routine analysis of drill cuttings and cores, both as whole specimens and as processed samples (e.g. thin sections).

The actual categories of data were to include:

- basic petrology for all classes of rock (particles, matrix, cement, and voids);
- pore fluids (especially hydrocarbon shows);
- palaeontology (either general or specific);
- structures (where visible, including fractures).

### Interval versus point data

In a bore hole or core, which essentially represents a one-dimensional traverse of the stratigraphy, data are related to depth in two ways:

- as intervals, between two bounding depths, in which certain features are observed;
- as points, at which spot samples are taken or measurements made. A point may be a core plug or chip measurable in millimetres.

### Distribution

Geological data intervals may be entirely regular, in which case they are referred to as incremental or irregular and non-incremental. Incremental data are normally derived from material such as accumulated drill-cuttings, which are sampled and described in bulk from regular intervals (e.g. 1 or 5 m). This increment may change between hole sizes, beyond certain depths, or upon penetrating key zones. To describe cuttings speedily, any practical system must work on predetermined intervals and yet retain the flexibility to change them on-line as circumstances dictate.

Non-incremental data reflect either the true nature of the geological record or the sporadic nature of the sampling. While it is possible to describe a core at even increments of, for example, 25 cm, it is unlikely that such a description will be as valuable as one based on the true boundaries of features. Non-incremental data may be intervals of irregular thickness, or irregular distributions of points. All these types of data should be members of the same population of data.

### Detail

Even at the mesoscopic level the descriptions attached to the distributions will vary considerably in their detail. A lithology as yet undrilled may simply be a name distributed between two vertical depths or time markers. An observed mineral or secondary lithology may be defined in terms of composition, texture, and a distribution both against depth and within its host. These relationships are highly diagnostic and must be preserved definitively by a model.

## Imprecise terminology

If describers are not sure of the descriptive vocabulary they may restrict their descriptions, or invent synonyms. For example, the American Geological Institute defines the term 'sucrosic' as completely synonymous with 'saccharoidal' and 'sugary' (Bates & Jackson 1987), while the American Association of Petroleum Geologists' (AAPG) widely adopted list of abbreviations includes all three terms (Swanson 1985). This may confuse the inexperienced geologist, who might be tempted to use all three terms, either interchangeably or by constructing an arbitrary local scheme to differentiate them. This hinders later manipulation of the data, despite there being an underlying true definition of this texture.

Definitions of an object may vary according to the scheme of classification for which the nomenclature is developed. A carbonate rock might be correctly classified according to rock-framework (Dunham 1962), particle and cement type and abundance (Folk 1959), or grain size (Grabau 1911). In order to use these schemes logically, the underlying definitions must be determined, any common areas identified, and a nomenclature imposed on the whole. Where there is no overlap users will normally choose to use one scheme or another.

In the absence of standardization each company, and even parts of companies, may have their own nomenclatures. Those who have chosen to instill some rigour into the procedures remain in the minority.

## Human variability

Geological data relating to the same object are collected at a number of stages by various individuals. With each subsequent description of a sample the quantity and quality of the data will vary with the expertise of the describer and the time available. Furthermore, each part of the process will be carried out with a bias to the perceived application of the data. At the basic level of description, a mudlogger, who may not always be a geologist, may describe a rock as 'igneous' for the mud log. A general geologist may improve this description to 'granite' for the lithology log, while a specialist igneous petrographer with the benefit of thin sections may classify it as 'monzo-granite'. Each is correct within the limits of experience and immediate application.

## Linguistic variability

Given the international nature of geological prospecting, is the possibility that the nationality of the providers and the users of the data may be different. Although the main working language of the oil industry is English, many of those who work in it speak English as a second or third language. Despite high levels of geological proficiency, the quality or detail of the description may be compromised if the describer is not working in their preferred tongue. Frequent recourse to a dictionary or list of abbreviations hampers freedom of thought, and speed of description. In practice, even the best workers may use only a fraction of the available vocabulary, and restricted descriptions become the norm. Practically the choice may be between good language skills and good geology.

## The geological descriptive process

The wellsite geological process was functionally decomposed to identify the critical issues for real-time description. In the conventional computerized process the situation is as follows: the geologist is looking down the microscope, trying to remember which features to describe, the vocabulary to use, the correct (or even the last-used) abbreviation, and the order of written description, remembering to check for, and describe, hydrocarbon shows, and writing all this down on a worksheet; then, when time is available, or a report is needed, transcribing the notes (including those of others), and entering the details of symbols with text in the correct order onto a computer plot package; then manually proofing it. In the course of this process detail is sacrificed to time and plotting space. Twice or three times a day the individuals may change shift, and a generalist may be replaced by a specialist: the phenomenon known in the oil industry as the 'tourly unconformity'.

Traditional practices were scrutinized. Is it, for instance, valid to describe cuttings samples of one lithology as 'green to yellow', 'firm, becoming soft', or 'occasionally red'? These practices result in a degradation in the quantity and quality of data available to end-users. This is largely a result of the plot-centred approach, where the only product is a compressed-scale plot, e.g. the mudlog or the composite log, where descriptive detail is limited by space and aesthetics. If the data are not on the log they are lost.

## Confidence: known versus unknown

In any sample there may be features which are not fully identifiable. Geological features are often parts of a spectrum of subtle change, the confident interpretation of which are critical to later interpretation. In established wellsite practice there is little room for the unknown. As a result, if a mineral is white and powdery it may appear on a log as 'anhydrite', even if the observer has insufficient experience to distinguish anhydrite from kaolinite or rock-flour; since all three will disperse in water. Rarely will the describer admit uncertainty; the unsure usually omit to describe the feature.

In designing a system to maximize accessibility and confidence we felt it necessary to encourage as much description as possible, but to discourage low-confidence attribution. Where an observer is unsure it is better, in the interim, to allow an unknown mineral to be described in detail. In the example above it would be 'unknown: white, powdery, soft'. We decided that this was more useful to the end-users and second stage petrologists than a name which might mislead interpretation.

## Confidence: real versus inferred

The converse of the situation where the geologist can see but cannot name is that where he or she can infer but cannot see. While drilling a well a mineral or lithology may be absent from cuttings samples owing to dissolution or dispersion but evident from other data. Common examples are soluble evaporites and some swelling clays. The feature must be named in order to be included in the model and any subsequent interpretation; but is assigned the status 'inferred'. The description may be expanded as more data become available (e.g. from sidewall cores) and the status changed to 'real'.

## Areas for potential improvement

Our analysis of data flow, confidence, and degradation, led us to identify areas of possible improvement. These were to:

- improve quality and geological correctness of descriptions;
- ensure consistency of descriptions between workers;
- allow any geologist to work in their preferred language;
- increase detail without disproportionately increasing labour but also to optimize ease of use without sacrificing access to detail;
- make the entry of text and symbols, and any digital data (like grain size) as easy as possible;
- make the data fully editable;
- build the potential internal logic to link attributes and trap geological contradictions;
- produce a range of logs and reports with wide variation in data density;
- ensure that final descriptions and logs were entirely consistent with a user's specification, without increasing workload;
- ensure that data were never sacrificed to output capability;
- facilitate ease of data exchange from one database to another;
- develop a model independent of specific platforms.

## Logical development of the model

The critical issue was not the user interface, or the final plot formats, but the efficiency of the data model. Once an efficient model had been established then the other processes would follow naturally, provided that their needs were anticipated. In simple terms the model describes rocks as fragments (clasts or crystals), voids, and interstitial material which may be crystalline or particulate.

In an ideal model the geologist would record not names but observations. The data might consist, for example, of:

- clast or crystal size distribution (if any);
- matrix percentage (if any);
- cement percentage (if any);
- chemical or physical composition and texture of all the above;
- any artefacts or non-constituent accessories;
- colours associated with all the above;
- voids, character, abundance, connectivity, and contents.

With these data stored the user could determine the sample attribution under any suitable scheme of classification (not contextual or depositional). For instance, siliciclastic rock could be classified according to percentage of each particle size class and supporting material. These precise values could be mapped multi-axially and a scheme superimposed on the distribution (e.g. a Folk, Piccard or Pettijohn ternary grid) to produce definitive names. Values such as sand sorting coefficients of the grain fraction could be added to refine the classification.

In practice it is not always possible to perform a full analysis of particle size distribution. At the well-site the fastest methods involve the selection of general names; without these the process would become intolerably slow. It is possible to preserve each selected name as a set of grid coordinates to facilitate later reclassification according to different schemes, without implying detailed analysis and compromising data confidence; as storing median compositional values might.

The foundation designed around the realistic ideal can accommodate detailed laboratory descriptions and analyses (e.g. thin-section data) as well as more cursory observations.

## The database structure

### The geological data model

The geological data model described is complex in terms of relational database design. There is no simple hierarchy of data structures and properties that can be represented easily, which leads to relational anomalies such as:

- rocks such as conglomerates may include components, which are in themselves rocks;
- structures and hydrocarbon shows which may cross lithological boundaries;
- when plotting related geological data, there should be a way of associating composite symbols with those data.

The logical model was defined in terms of traditional entity-relationship diagrams. Figure 1 shows an overview of that part of the model which deals with petrological and structural data. At the top of the diagram is the 'hole' entity, which acts as the connection point to the rest of the wellsite data model, in which are described the many engineering and physical attributes of drilled well paths. Below 'hole' is 'geology', which defines an interval over which geological data are stored. This may correspond to a section of core, a sidewall core, or a sample of cuttings. This entity then has a one-to-many relationship with a number of other entities, including:

- GEOCHEM: storage of the results of geochemical analyses;
- FRACTURE: type, spatial distribution, and description of fractures;
- CALCIMETRY: calcite and dolomite fractions of the sample;
- SHOW: hydrocarbon show data;
- ROCK: linked to more detailed lithological descriptions;
- STRUCTURE: links to different structure types (sedimentary, deformational, and organic).

Most of the detail describing lithological samples is contained in the entities below the 'rock' entity. 'Rock' is subdivided into one or more 'lithology' items, and each of these may have one or more instances of the following properties:

- POROSITY: type and amount of porosity;
- COMPONENT: distribution, description and amount of lithological components;
- COLOUR: distribution and description of colouration;
- LITH_OUTPUT: how the lithology is to be represented on plots.

This model provides the framework for storage of geological data. To populate the data items within the model there are two catalogues within the database. One catalogue contains the definitive master dictionary of geological terms or archetypes; the other provides the ability to link stored attributes to symbols for plotting. It is these catalogues that provide the links between the uniquely defined attributes and the users' own nomenclature and preferences. It is the structured digital nature of the catalogue that allows for the application of internal logic.

### Implementation

The model is currently implemented on both relational and object-oriented database systems. To date it has been used to store wellsite geological descriptions of core, cuttings, chips, and other samples. The data model is reasonably compact, and lends itself to being scaled down to run on almost any modern computer architecture.

### Data storage and retrieval

To give applications programs access to the geological data, a set of C language functions was been created to manipulate geological data. The load, save, delete, insert, and update

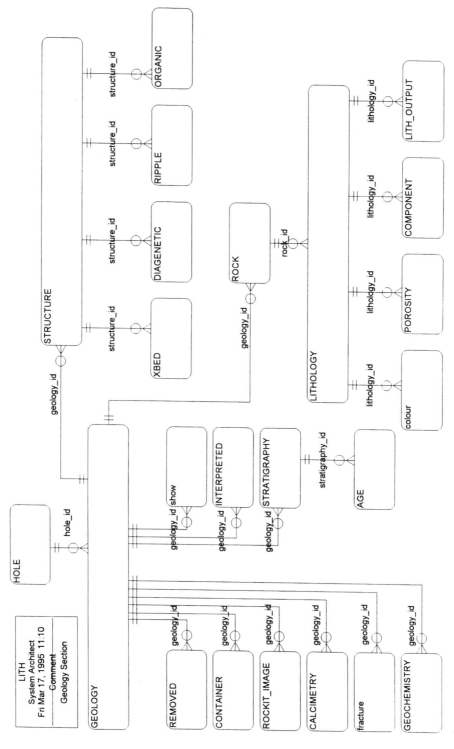

**Fig. 1.** An overview of part of the model that deals with petrological and structural data.

functions have been implemented, along with functions to set and change individual attributes of data structures. Geological data are retrieved in structures that closely resemble the database datamodel.

## Master dictionary

We developed a comprehensive, synonym-free dictionary of attributes to populate the model. When completed for field release it comprised over 2000 defined items, covering hydrocarbon and mineral exploration, sedimentary, igneous and metamorphic rocks, and their compositions, textures, fabrics and structures. Of these, any given user would probably use only a small proportion. The majority of the attributes are uniquely definable, based on the expertise of numerous source documents (IUGS 1973; Bates & Jackson 1987). The unique definitions allow for the confident translation of data, yet the dictionary remains flexible enough to accept any new, uniquely definable, items.

The storage of colour attributes demonstrates a typical challenge. Colour in visible light is a spectrum of continuous data, notorious for its descriptive imprecision. There are a number of ways to define and store a colour: it can be named (e.g. salmon pink); stored as red, green, and blue (RGB) intensities (e.g. 202136178); as Munsell's hue, lightness, and saturation (e.g. 5R 7/4); as a cyan, yellow, magenta and black (CMYK) value, or even as a printer's Pantone number (e.g. PMS 288). All of these are easily definable except the name, yet it is the name which usually appears in written descriptions. It is possible to present thousands of colours on a screen; but there are limitations to the veracity of computer colour-mapping, and problems in comparing transmitted to reflected colours. It is also difficult to arbitrarily select and name parts of a continuous spectrum. The most effective solution was to take an accepted classification system, in this case the Geological Society of America's (GSA) *Rock Color Chart* (Geological Society of America 1991), which presents paint tiles for each of a number of Munsell values and gives them names which can easily be replaced. This approach allowed us to establish the exact perception of, for example, buff or tan or lavender for each user dictionary while retaining the possibility of future conversion to other systems (e.g. RGB).

The dictionary was compared with the legends of various major oil companies and certain standard works (Swanson 1985). Problems of missing but definable attributes were resolved; shortcomings in the standard literature (principally duplication and omission) were recognized and avoided wherever possible.

The approach with the dictionary was to:

- identify discrete data, and ensure that they could be incorporated;
- identify continuous data, and then establish the most efficient way to render them usable.

## Customization: user dictionaries

The establishment of the master dictionary of archetypes laid the foundation for the creation of user dictionaries. Overlying each implementation of the model is the master dictionary in which definitions are held; over the master dictionary lie the user dictionaries. Any interaction between the user and the model takes place through a user dictionary. The user dictionary establishes correspondences between defined archetypes and local nomenclature. For instance, the archetype for grain size 15.6 to 31.2 µm may be linked to 'medium silt' for BP and 'Silt grossolana' (coarse silt) for Agip. Each will see only their own terminology but can confidently translate descriptions in the knowledge that the size category is preserved. This can also apply to database mnemonics. The user level also includes data ordering and symbolic dictionaries, to ensure as near perfect descriptive text as is possible.

## User preferences

As a replication of the master dictionary, the user dictionary contains too many options for any given well or part of a well. Above the user dictionaries are two levels of user preference: global and local. Global preferences deny or enable access to any given attribute by any worker using the user dictionary (e.g. corpohuminite, or talc, on an oil-well). Local preferences are set by the field worker to restrict the choices presented temporarily to those most likely to be needed; they can be unset instantly. Neither set of preferences affect the retrieval of stored data; they are merely windows through which access is given, local users cannot define terminology.

## Language customization

A natural consequence of the user dictionary system is that alternate languages can be used to access the data. The use of translation dictionaries at the user level allows different workers in the same network to access data in mutually incomprehensible languages, and potentially different scripts. The use of different languages presents a higher level of difficulty; it is essential to be confident that geological terminology is translated, and not just transliterated.

## User interface

### Graphical user interface

In order to enter the observed geological attributes as rapidly and confidently as possible, a suite of programs was created to act as the graphical user interface (GUI). Each program is specifically suited to accessing different forms of data (e.g. cuttings or core) rapidly.

Once a user is established as belonging to a recognized user group, then all the choices presented through the GUI use the language and terminology specified for that user group. The GUI presents those choices currently enabled by the user preferences. An example main GUI for cuttings description is shown in Fig. 2.

There are specific screens for the entry of different data types and sets of related attributes, as well as particular aspects of data manipulation. The second example interface Fig. 3 shows a graphical screen that assists in the creation of an interpreted lithology record from the percentage cuttings lithology. Reference curves such as gamma-ray or drill rate may be displayed alongside lithological data during the process to improve the picking of bed boundaries. The selection of boundaries is performed by direct mouse interaction with the graphical screen rulers.

### Managing complexity

To keep the implementation of the system practical, levels of descriptive detail were established which were felt to be consistent with the quantity of data available at the wellsite. For example, a limit of five was imposed, via the interface, on the number of lithologies within a cuttings sample and on the number of cements or matrix types subsidiary to each lithology.

### Procedural guidance in the GUI

The GUI layout, as well as program logic, guides the user through the descriptive process;

the show description process, for instance, is laid out from left to right on one panel.

To increasing the objectivity of descriptions, wherever possible attributes are represented symbolically on the screen, or supported by auditable external aids (e.g. GSA rock colour chart). Selection of items such as grain morphology is accomplished with the aid of glyphs, representing sphericity and roundness.

### Geological objectivity

The validity of the system depends upon a clear understanding, by the user, of the definitions chosen. An on-line geological 'help dictionary' to which users can refer, in addition to the on-line system manual, guides users through the system and reduces subjectivity. This help dictionary is also used by administrators setting up user dictionaries. At the most basic level it is text-based, containing not only definitions but also general information on procedures and practices (for instance, how an acetone test is used to reveal gas condensate shows). At a higher level it may contain graphics (e.g. bedforms) or photographs (e.g. mineral thin-sections in different states of polarization). This information, which is both corroborative and educational, is an essential part of the attempt to improve overall confidence in the accuracy of the data.

### Customizing the user interface

Only active items will be presented on the GUI, each individual user can remove active items from choice lists in order to speed selection. The order in which items appear on lists can also be changed: the internal ordering is geological, and lists will normally be displayed in this form (e.g. all igneous rocks together in IUGS order) which allows for rapid access via speed-keys.

Through the internal logic the user is either not presented with items inconsistent with previous selections, or can be warned of potential conflict; for instance the selection of a carbonate rock will initiate a series of options consistent with the description of carbonate rocks; the user group settings would govern (for instance) whether that selection was from Dunham (1962) or Folk (1959) classifications. In addition to procedural guidance, logical checking, and trapping of contradictions, in suitable cases the selection of one attribute may lead to the automated storage of inseparable attributes (e.g. a user selecting shale might find that the

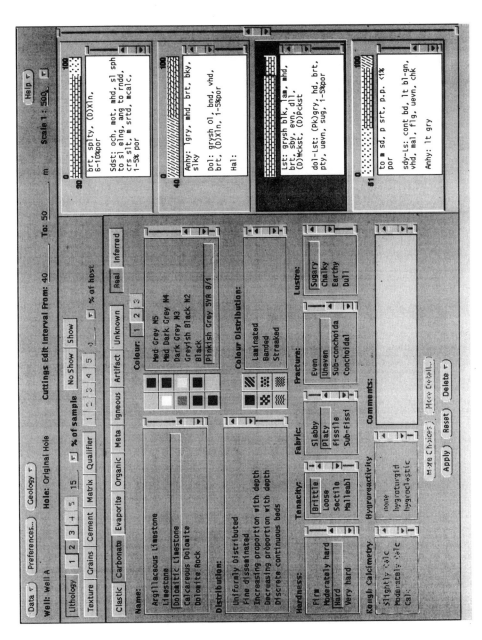

**Fig. 2.** An example main GUI for cuttings description.

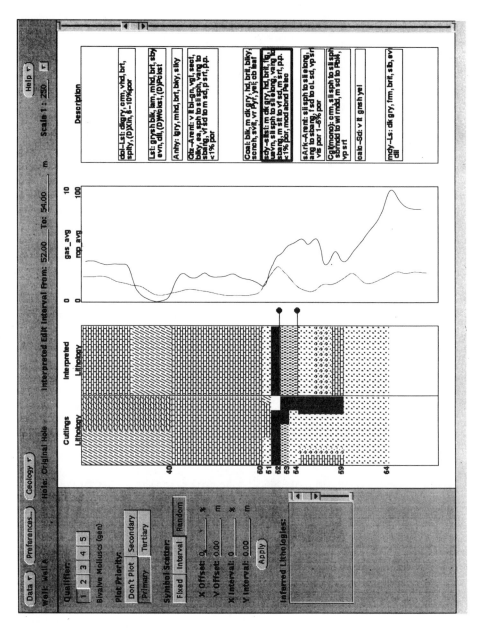

**Fig. 3.** The second example interface.

attribute fissile is automatically stored). Procedures can be made compulsory, for instance hydrocarbon show assessment in selected reservoir type lithologies.

## Data usage

Production of plots plotting from such a wealth of defined data requires a data-centred approach. The plot is no longer the beginning, middle and end of the process; it is just one of many ways of viewing all or part of the contents of the database in its most recent state. Data can be plotted and replotted as new information becomes available, or when new formats or symbols are defined.

The data access functions allow for simple extraction of geological data from the database. The geological structures returned from the database contain the symbols that are needed to plot them, along with other data items that define the positioning and frequency of accessory symbols and modifiers.

### Text reports

The ease of extraction of data structures, combined with the user dictionaries, give the user of the system the ability to create reports in standard terminology, in a set order, in a desired language, and in a long or short form, even though the data may have been stored under a completely different set of conditions. By setting up templates exactly as much information as is required can be extracted and printed (e.g. report only the first two recorded grain minerals).

### Integration with other data

The geological model is a small part of the much larger wellbore data model. As such it stores data in parallel with a variety of other oilfield data, such as depth-based drilling data, gas and hydrocarbon data, wireline and MWD data. This enables the production of composite logs and reports that call upon any of these data. Data from other sources may also be imported into the database; or geological data may be easily exported to other databases or spreadsheets.

### Data transfer

Careful attention was given to the problem of translating the definition of recorded features into other existing defined systems (e.g. database mnemonics) and other formats for geological data transfer. For example, it is possible to extract the data from the database in WITS (Wellsite Information Transfer Specification), pre-defined records for wellsite geological and hydrocarbon show data (Rose 1991). The most basic level of transfer between environments is ASCII text; the highest is a transmigration of the model and transmission of native data.

## Advantages of the data-centred approach

### Describer benefits

The describer, guided by the interface, enters the observations directly into the database by selecting easily recognizable attributes in any order. The text description will be correctly spelt in the correct order for that project, correct symbols are automatically selected, and any additions or revisions will be inserted in the correct place in the text.

When in doubt, the user can browse the help dictionary. This will help to improve the depth and consistency of descriptions and analytical methods. This is a significant step towards gathering more valid wellsite data prior to laboratory verification. If for any reason the describer feels that the model is inadequate to describe what is seen, there is a free-form read/write field for comments attached to each lithology; entries in which inevitably fall beyond the reach of the internal logic.

All that remains for the describer to do is assign, where necessary, priorities to the various features and attributes so that the most significant data are shown on condensed plots. If output is required in other users' languages or symbols the change can be made at the touch of a button. All this is performed in the certain knowledge that no data have been lost or corrupted during any of the reporting processes. If data transmission is required and transfer protocols have been established, then the communication should be effortless.

### End-user benefits

The immediate benefit for the end-user of the data will be improved quality and consistency. The descriptions are, probably for the first time, entirely consistent with previously agreed classification systems, abbreviations, descriptive order, layout, and degree of stored detail; and everything should remain consistent for the duration of the well.

Corrections can be made at any stage, and extra observed detail can be stored without compromising the brevity of summary plots and reports.

Individual items or whole groups of data can be selectively queried, retrieved and displayed in a manner that would previously have required manual re-entry of the data. For example, grain size and simple grain size distribution can be plotted against other data, including petrophysics; products might be gamma-density versus percentage cementation, resistivity/grain size cross-plots, or graphic core logs.

This facility makes the geological descriptive data accessible for the first time to other parts of the geoscience team. Although descriptive data might still be regarded as subjective by many petrophysicists and reservoir engineers, the system can improve the confidence level of the data for all users. For users with a petrological database of their own, the benefits are even greater; with established protocols the data can be transmigrated to the highest level of stored detail on a regular basis. Thus the wellsite data are made continuously available to operations and production staff in digital form throughout the drilling and evaluation phases of the project.

## Trading data

Where projects are sponsored by partners with different internal data standards it should be possible to present the data to each in their preferred format, and even, in the case of national authorities, local language.

## The future

The system has been subjected to rigorous use by both internal and external client geologists. It is currently implemented in more than 50 mudlogging laboratories around the world. Translations into Norwegian, Spanish and Italian have been undertaken, and the descriptive preferences of various companies have been set up.

The future of the system must be driven by the needs of the users. The interfaces will continue to evolve to further mimic the geological descriptive process. It is probable that graphic and symbolic systems will predominate as hardware capability improves; this will also include the use of photographs. Eventually voice-input may allow data to be entered without the describer's eyes ever leaving the microscope; and digital cameras may record colour, lustre, texture and morphology, with the images themselves becoming part of the record. In the absence of precise industry-accepted standards for the exchange of detailed geological data, a protocol will have to be established. Various hard-copy formats will need to be designed to gain maximum value from the data. The advisory nature of the system can be enhanced to the point where identification, to degrees of confidence, can be suggested to a describer once attributes have been entered. It will only be possible to perform these actions with confidence because the data are so highly defined.

This original version of this paper was first presented as SPE 27542, Digitizing Rocks?: Standardizing the process of Geological Description Using Workstations at the 1994 European Petroleum Computer Conference, Aberdeen, 15–17 March 1994; and is reproduced here in a revised form with their kind permission. The authors are also grateful for the permission and support of Baker Hughes Inc., the comments of the internal reviewers, R. Burman and P. O'Shea, and the invaluable critique by J. Laxton.

Credit is given to BP Exploration Operating Company Limited and Den Norske Stats Oljeselskap a.s. (STATOIL) for their contribution as partners in the slim-hole coring venture that drove the development of the system. Particular credit within Baker Hughes INTEQ must go to P. Morrison and M. Ridgway for programming the geological and functional specifications on the interface and database respectively. L. Marks tested the code rigorously from an engineer's perspective. The authors are also pleased to acknowledge the contributions made by P. Spicer and A. Gray of BP XFI, and M. Darke and P. Geerlings during the first field test. We would also like to thank the geoscientists whose expertise underlies the dictionary.

## References

BATES, R. L. & JACKSON, J. A. 1987. *Glossary of Geology*, 3rd edn.

DUNHAM, R. J. 1962. Classification of carbonate rocks according to depositional texture. *American Association of Petroleum Geologists, Memoir*, **1**, 106–121.

FOLK, R. L. 1954. The distinction between grain size and mineral composition in sedimentary-rock nomenclature. *Journal of Geology*, **62**, 344–359.

——1959. Practical petrographic classification of limestones. *AAPG Bulletin*, **43**, 1–38.

GEOLOGICAL SOCIETY OF AMERICA 1991. *Rock Color Chart*.

GRABAU, A. W. 1911. On the classification of sand grains. *Science*, **33**, 1005–1007.

INTERNATIONAL UNION OF GEOLOGICAL SCIENCES, Subcommission on the Systematics of Igneous Rocks 1973. Plutonic rocks, classification and nomenclature. *Geotimes*, **18/10**, 26–30.

JANTZEN, R. E., SYRSTAD, S. O., TAYLOR, M. R., SPICER, P. J., STOCKDEN, I., DODSON, T. & SAUNDERS, M. R. 1993. *Answering Geological Questions from Slimhole Coring Exploration.* Paper T, SPWLA 34th Annual Logging Symposium, June 13–16, Calgary, Alberta, Canada.

MORLEY, A. R. 1990. *Workstations for the Well site: Using New Computer Standards to Implement an Integrated Information Management System for Drillers, Engineers, and Geologists.* SPE 20330, 5th SPE Petroleum Computer Conference, June 25–28, Denver, CO, USA.

PETTIJOHN, F. J. 1954. Classification of sandstones, *Journal of Geology*, **62**, 360–365.

ROSE, R. J. (ed.) 1991. *Wellsite Information Transfer Specification*, RP 3855, American Petroleum Institute.

SPAIN, D. R., MORRIS, S. A. & PENN, J. P. 1991. *Automated Geological Evaluation of Continuous Slim Hole Cores.* SPE paper 23577.

SWANSON, R. G. 1985. *Sample Examination Manual.* American Association of Petroleum Geologists.

WENTWORTH, C. K. 1922. A scale of grade and class terms for clastic sediments. *Journal of Geology*, **30**, 377–392.

# Towards the creation of an international database of palaeontology

## M. LHOTAK & M. C. BOULTER

*Palaeobiology Research Unit, University of East London, Romford Road,
London E15 4LZ, UK*

**Abstract:** The Plant Fossil Record (PFR) database is being built to include biological, geological, geographical and bibliographic information of all authoritative plant fossil citations throughout the world. The interactions between these have created standardized records from different sources and programs have been written to add and reorganize data. Techniques such as optical character recognition (OCR), e-mail and diversity interfaces are used. Interdisciplinary and cultural aspects of this ambitious project are being considered. The work is to help as many different specialists as possible to use palaeontological data and to relate these data to evolutionary and biostratigraphic problems as well as to climatic aspects of global change. The structure of the database comprises some original features. It is a temporal facility: each edit or addition is timed and dated so that its state at any particular time can be recalled. Geological time and the associated global palaeogeography are also built into the database giving a time dimension to record likely evolutionary trends.

The Plant Fossil Record (PFR) database is organized through the International Organisation of Palaeobotany. The first version, PFR1, became available in 1991. This contains 10 477 taxonomic details of most plant fossil genera and some of the records have stratigraphic and morphological descriptions incorporated. The database has now been redesigned to include individual fossil occurrences together with geographical and stratigraphic details. This paper explains the reasoning which led to the new design, PFR2, and describes some features of the database which are unique to this second version.

The first conceptions of the database were strongly influenced by the ideas of Hughes (1989) who has argued that binomial nomenclature hides a lot of the information that is available in individual fossil biorecords, or palaeotaxa. From these beginnings, the style of the descriptive part of PFR was determined by an international workshop of about 75 palaeobotanists who agreed the 'Frankfurt Declaration' in 1990. This established the 43 field titles which are still used in the generic description part of the database (Table 1).

The structure of the database comprises some novel approaches to data modelling which reflect the heterogeneity of the data sources. These features include data normalization, identity of data records, hierarchisation of data, synonym processing, spatial data structures and temporal facilities. The resulting model (Fig. 1) and the cultural differences in both the data records and the users, result in the need for the construction of a new human-computer interface which can provide various forms of remote access to the data. In order that the database can be of the most use to as many people as possible it is important that this international facility produces without delay a large working model for the entire scientific community. The database may be of use in helping to solve problems connected with changes in climate and environment, rates of extinction and evolution, biostratigraphy, and our general understanding of biodiversity (Boulter & Fisher 1994).

## Sources and nature of the data

We began by taking some of the magnetic data of taxonomic authority that existed in 1990 and adapting them to the PFR1 standard format (Table 1). For the descriptive taxonomic records of the PFR1, the fossils from a computerized version of *Index Nominum Genericorum* were retrieved. This index is edited from the Smithsonian Institution, Washington, DC, and attempts to give bibliographic details of all plant genera as well as other nomenclatural information. Details from each of the genera so captured were checked manually (mostly by P. Holmes, A. Hemsley and I. Poole) against the printed catalogues of Andrews (1970), Watt (1982) and Blazer (1975) and additional taxa were included from the more recently published generic index of Meyen (1991). To help search effectively, geographical, stratigraphical and botanical jargon should probably be standardized with the same spellings. Some progress was made to institute a standard (Holmes & Hemsley 1990) and this has been incorporated into the PFR1. PFR1, made available in 1992, was attached to a friendly database manager called Textmaster.

*From* Giles, J. R. A. (ed.) 1995, *Geological Data Management*,
Geological Society Special Publication No 97, pp. 55–64.

**Table 1.** *PFR1 fields as agreed in the 'Frankfurt Declaration'*

1. Name of genus
2. PFR number
3. NCU status
4. Organ
5. Type species
6. Author of generic name
7. Date of publication
8. Author(s) of article
9. Year of publication
10. Title of article
11. Journal or book title
12. Journal number
13. Page number
14. Basionym
15. Synonymy
16. Typification
17. Attach. or assoc. organ
18. Botanical rank order, or
19. Botanical ranking family
20. Diagnosis
21. Description
22. Preservation
23. Comparison record present
24. Region
25. Locality
26. Facies
27. Sedimentary basin
28. Lithostratigraphy
29. Sample position
30. Sample lithology
31. Era
32. Age of sequence
33. Radiometric age
34. Biostratigraphy
35. Repository
36. Links to quantitative data
37. Links to palynodata
38. Links to ING
39. Links to bibliography
40. Links to more details
41. Advisor/date added to PFR
42. Source reference for PFR
43. Other notes

There is no more systematic and taxonomic data available in a magnetic medium and so further versions of PFR will have to be created manually or by optical character recognition from printed literature. This will be time consuming and expensive, but should be realistically attainable if the international palaeobotanical community can be inspired to become involved. PFR2 already includes modern genera which have fossil species as well as records of plant fossil occurrences (PFO). The bibliographic data is being constructed from magnetic versions of Tralau's (1974 and 1983)

fossil plant bibliography as well as from the other magnetic sources mentioned here. At some future date it may be possible to organize the inclusion of annual regional bibliographies which are prepared regularly by many palaeobotanical organizations.

PFR2 includes plant fossil occurrences and has the potential of being most useful to a variety of environmental and evolutionary scientists. Each record contains nomenclatural, geographical and stratigraphical details within the 18 fields whose titles are listed in Table 2. The first major source of this occurrence data is from the palynological database of Ravn (1989), which contributed thousands of records. There are numerous sources of additional records for the plant fossil occurrences. One major source, not necessarily published, are from the computerized catalogues of important museum collections which are now becoming available. Most of the records of specimens listed in these catalogues have geographical and geological information as well as the name of the collector. They are included in the occurrence database if they carry the authority of a named collector or author in a named museum or scientific journal.

## Processing the data

Standardization of the biological, geological, geographical and bibliographic language is essential to facilitate meaningful searching. Different values of the different initiating palaeontologists mean the languages used in the sources of the raw data are very diverse, but they can be and are being standardized. Some of the guiding principles are outlined below. Large databases such as the PFR are very good at giving easy access to information, but do not discriminate about the quality of the data. Some of the records are useful and reliable, others are of no use and others are incorrect; many fall between these extremes. Also, different authors and users have different ideas about which of these categories applies. The quality of work by palaeontologists will continue to vary so these problems will remain. The new situation that has been created by the database makes much more information easily available, so there will be more decisions to be made. However, easy access to large amounts of data enables efficient statistical comparison of occurences, so the inaccurate or wrong stratigraphical or geographical information can often be quite clearly detected on a graphical display or printout.

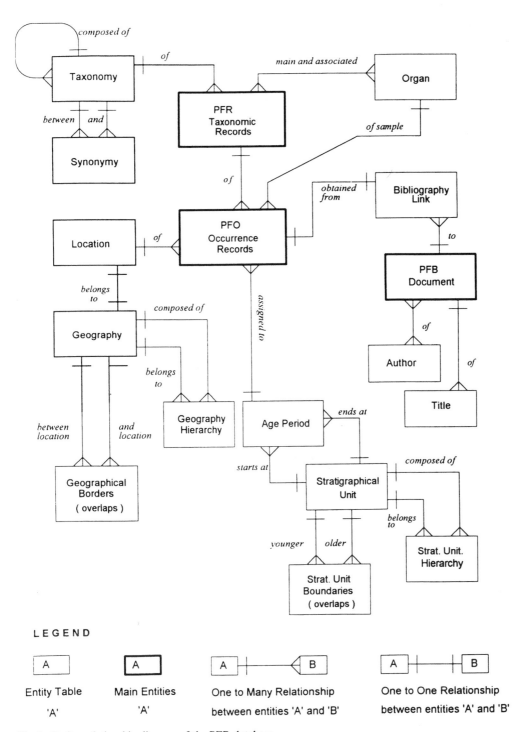

**Fig. 1.** Entity–relationship diagram of the PFR database.

**Table 2.** *The fields used in the PFO subset of PFR2*

1. PFR Number
2. Family
3. Genus
4. Species
5. Organ
6. Region
7. Location
8. Latitude
9. Longitude
10. Era from
11. Period from
12. Era to
13. Period to
14. Author
15. Year of publication
16. Source
17. Editor
18. Comment

## The import process

The first stage of every import process of a given dataset is to transform it into a computer-readable format. In our database we use ASCII flat files as a standard when importing new records. The unified format simplifies the import process and also increases the range of compatible data sources. If the source of new records is an electronic database, an export function from within the source database management system (DBMS) can be used to introduce all the entries into a single (or multiple) flat ASCII file(s). Such a function is referred to as a report generator. These flat files can then be directly fed into the destination DBMS. The import module of the PFR2 recognizes records and fields within the flat file and re-assembles appropriate data structures. Those records are not inserted into the main database directly; they are stored initially in a temporary database.

If the source flat file was created from an electronic document (such as an e-mail message or bibliographic record) another process must precede the importing process. This is required because of the absence of a structure of fields within the imported document. Often, a human expert has to apply their cognitive capabilities and mark blocks of text and assign the links to database fields. Another possibility would be to use a promising technique from artificial intelligence which could extract the information from plain text, but such techniques are in their infancy. Techniques using artificial intelligence would be most useful to process data in non-electronic forms, such as printed, typed and hand-written documents, because of the availability of the advanced technique of OCR. We have successfully (up to 99%) tested OCR to extract data from many different kinds of printed catalogues.

## Data cleaning

New datasets may contain misspellings and records might be incomplete or changed in some way. This often happens because of the corruption of discs through their being physically damaged, so, the next process is to clean the newly inserted data records. The first stage of the cleaning process is to build up the data dictionaries of the temporary database in which the newly imported data records have been stored. The new data dictionaries are then compared with the comprehensive data dictionaries of PFR2. The intersection of those obviously contains clean and correct data values (assuming that the main data dictionaries have not been previously polluted by an unauthorized user). The rest of the newly built data dictionaries must be reviewed manually by an expert, if such a person is available. If not, the origin of data is investigated and if that is not possible (for example, the author is unreachable, or the records are too old) the data can be marked as 'unclean' and will always be considered as unreliable and to be used with caution.

Another option currently being evaluated is the use of weighted attributes, as proposed by Zadeh (1972) in his *Theory of Fuzzy Systems*. This uses the theory of combinations to express the vagueness of terms used in human language. An attribute of a record in a database stores not only the value, but also the possible interpretation of that value, i.e. the most probable range, the less probable range, etc. to the definitively not possible range of values of a particular attribute. Once the data are ready to be inserted, the records from the temporary database are appended into the main PFR2 database with additional temporal pieces of information referring to the date of insertion. The system will also make sure that no duplicates are added to the resulting database.

Accurately copied data are accepted for incorporation into the main database without the cleaning process only if they originate from one of two sources. Either:

(a)  from a scientific journal that is normally acceptable to the scientific community-at-large as a source, or;

(b) from the specimen label or catalogue of a curating museum. The decision of validity (within the meaning of the International Code of Botanical Nomenclature) remains with the rules that have developed.

But the decision of acceptability for one particular record is with the data editor who creates new data entries.

## The structure of the database

The hierarchical structures (Fig. 1) are discussed below and can be understood as separate databases co-operating with the kernel database structured on the Frankfurt Declaration model (Table 1). The construction and content of these hierarchical synonymy data dictionaries is closely related to 'translating' records during the data-gathering process.

### (i) Taxonomy

The temporary data dictionary describes the system of classifying genera and is divided into three hierarchies: family, genus and species. However, a feature of most evolutionary biology is that these hierarchies are not always stable. This data dictionary has been constructed to allow easy reconfiguration of the hierarchies. If a hierarchy changes, a new hierarchy structure is generated and marked as primary. The database then uses this new hierarchical link as the preference and stores earlier versions. Another feature of the taxonomic data dictionaries enables synonymies to be identified. Once a synonymy link is defined, a new connection is created in the hierarchical tree of the data dictionary, so that the tree becomes a hierarchically oriented graph without cycles. However, there is one danger, associated with this flexibility: users can remap the links of the data dictionary to their needs, but other users may not be able to understand the reasoning behind the remapping. The solution to this problem is not to consider those remapped links on a global level, but to associate them with the particular local user environment, so the changes made by one user will not reflect on mapped links of another one.

### (ii) Stratigraphy

The stratigraphic terminology is taken from Harland et al. (1990) which offers an unambiguous range of names and datings for the relevant

fields. Both the names and dates may be controversial, and other authorities' words and numbers differ, but Harland et al. can be followed without difficulty. The mechanisms are implemented in the same way as for taxonomy, including hierarchies (era, sub-era, period, sub-period, epoch) and synonyms. Another dimension has been added, the time interval neighbourhood. This shows which time intervals overlap, follow or precede the given interval in both earlier and older sequences.

### (iii) Geography

The underlying principle is similar to the one used in stratigraphy dictionaries explained above. The hierarchy is obviously different: region, continent, country, geological plate, area, locality and exact location (expressed as longitude and latitude). We have attempted to follow the regions described by Hollis & Brummett (1992) as much as possible, so as to conform with databases being prepared for modern plant distributions, but their scheme does not include the oceans. The main problems to be considered and resolved within this data dictionary include not only events such as the political redrawing of borders, landscape changes and movements of continents, but also the scale of the location. While we can work quite well within the confines of localities of an area of a few square kilometres, searches over the broad areas of continents are often too fuzzy, since the area is represented by many locations rather than one exact point. Perhaps, each location entry should be considered only with a certain accuracy (or weight) which decreases as the size of the locality increases. So, the search only presents records which maximise the accuracy within a given query.

## The procedure for searching

The so-called 'Green Book' (Boulter et al. 1993) lists the names of taxa which are present in our datasets. Not all the 15 258 names in the book are present in PFR2 because we have deleted all names that occur once (many are wrongly spelt words). The process of selection of the names to be searched from the Green Book is one of two crucial stages in the whole process because it is here that the user must apply a knowledge of plant fossil names. For example, maps in Fig. 2 are of occurrences ascribed to the family Symplocaceae. All the names listed in Table 3 were selected from the Green Book for

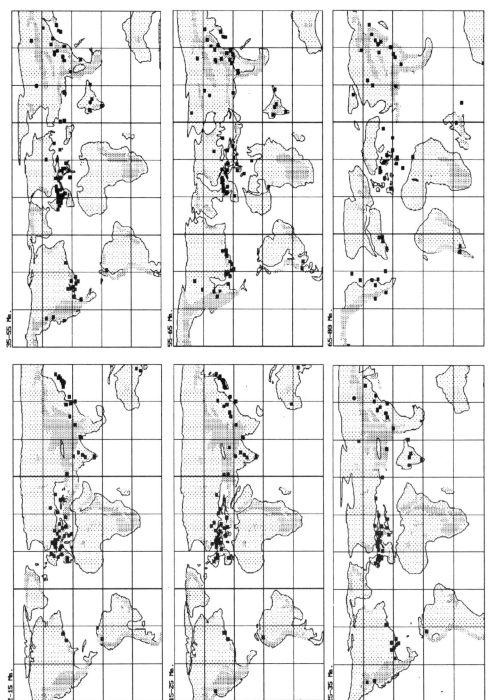

**Fig. 2.** Maps of occurrences of the family *Symplocaceae*.

**Table 3.** *Names from The Green Book selected to represent fossil occurrences of the Symplocaceae which we have searched from PFR2. Some of the names are valid formal taxa though of different rank, others are invalid, one has a spelling that is rarely used and another is morphographic*

Porocolpates, Porocolpopollenites, Symplocaceae, Symplocacites, Symplocarpus, Symploceaceen, Symplococites, Symplocoides, Symplocoidites, Symplocoipollenites, Symplocos, Symplocospollenites, Symplocoxylon

searching; the list includes common, rare, valid and invalid names of pollen, leaves and reproductive structures. The second crucial stage is when the user selects from the list of searched occurrence records those which are useful. It is the user that makes the decision, not a committee or controversial and difficult rulebook like the ICBN.

In practice searching works like this: from an e-mail station connected to the Internet the inquirer sends a message to boulter@uel.ac.uk, asking to be registered as a PFR2 user. A message is then returned to the enquirer, explaining how to log on to the database and how to make searches. The system is still at the developmental stage so do not be surprised if the response is slow or if the system is closed down. However, by summer 1994 the database should be connected to SuperJANET which will improve the response time of enquiries. Normally a

search within the 692 074 records of a taxon with about a hundred records takes a few seconds and the result should be immediately e-mailed to the originator of the search.

Table 4 lists a small sample of the results of the search for Table 3 taxa; those marked [M] are not plotted on the Fig. 2 because one user thinks they are unreliable. A different user may have different criteria of reliability, or different objectives in using the database, so may select a different set of data for mapping or using otherwise. Enquirers are expected to pay for the use of the facility by completing new occurrence records (a programme is sent free of charge to all users who register) of the fossils that interest them.

## Querying concepts

PFR1 implemented the Frankfurt Declaration model and includes all the data that were readily available two years ago. The PFR2 database is more about the representation of data and, as such, puts constraints on what is actually in the database. Consequently there are two possible ways of querying the database; as we will explain, the second one can only be implemented from within the PFR2 database framework. We call these two basic principles Query by Phrase and Query by Term.

Query by Phrase is perhaps the most common and simple query technique. The user asks to: 'Look in the database for a given word, words,

**Table 4.** *An example of the output from PFR2 from a request to search for 'Symplocaceae', of which there are 97 records.*

| | | | |
|---|---|---|---|
| 1027048 | CIS(WSL) [M] | 65.00 – 2.00 | MCHEDLISHVILI, N. D. 1964 |
| 1034672 | HUNGARY | 50.00 – 56.50 | KEDVES, M. 1961 |
| 1046895 | GERMANY | 35.00 – 56.00 | POTONIE, R. 1948 |
| 1051154 | HUNGARY | 35.00 – 56.00 | KEDVES, M. 1964 |
| 1088205 | CIS(KAZAKH ) | 35.00 – 56.00 | KUPRIYANOVA, L. A. 1960 |
| 1097795 | FRANCE | 61.00 – 65.00 | SITTLER, C. 1965 |
| *1144260 | INDIA(MADRAS) | 35.00 – 56.00 | VENKATACHALA, B. S. 1971 |
| 1175166 | CIS(KAZAKH ) | 60.50 – 65.00 | BLIAKHOVA, S. M. 1971 |
| 1182175 | EGYPT | 86.50 – 63.00 | KEDVES, M. 1971 |
| 1204374 | USA(ALABAMA) | 35.00 – 56.00 | TSCHUDY, R. H. 1973 |
| 1217296 | ITALY | 2.00 – 65.00 | LONA, F. 1963 |
| 1276726 | CIS(KAMCHATKA) | 23.00 – 65.00 | ZAKLINSKAYA, E. D. 1973 |
| 1277070 | ARGENTINA | 56.50 – 65.00 | ARCHANGELSKY, S. 1973 |
| *1287668 | CIS(VOLOGDA) | 50.00 – 56.50 | KUZNETSOVA, T. A. 1974 |
| 1290390 | INDIA | 2.00 – 65.00 | NAVALE, G. K. B. 1974 |
| 1404408 | WORLDWIDE [M] | 65.00 – 2.00 | MULLER, J. 1981 |
| 1451574 | EUROPE [M] | 35.00 – 56.00 | MULLER, J. 1974 |
| 1482979 | N.HEMISPHERE [M] | 23.00 – 65.00 | HOCHULI, P. A. 1979 |

*Records are rejected by one user and do not appear on the palaeogeographic maps.

combination of words, etc.' The database is scanned and everything that matches a precise query pattern is retrieved. For example, if we want to retrieve all records referring to 'Germany', then only those genera, where location is literally equal to 'Germany', will be retrieved. However, those locations referring to larger areas like 'Europe' or smaller areas like 'Berlin', 'Westphalia', etc. will not be retrieved because they do not match the pattern.

The solution to this weakness is represented by a different type of query, the Query by Term, which is only feasible with the support of the data dictionaries built into the PFR2 (as we have explained earlier). A query by a given term has first to construct the retrieval pattern set. The set will primarily contain the term itself and all successors of that term within a given hierarchy. The second step looks for synonymies of terms already in the pattern set and if any are found, they are added to the set as well. Then the process recursively repeats for each newly added term till the complete enclosure of the search set is constructed. The pattern set is then processed in a query and appropriate records are retrieved. For example, for a given term 'Germany', the records referring to 'Rhine' and 'Westphalia' are inserted into the pattern set, since they are lower in the hierarchy originating with 'Germany'.

Another promising type of query is induction retrieval. Its origin comes from the theory of artificial intelligence. The power of this type of retrieval is that we ask the same question iteratively: 'What is most remarkable in the contents of the database, except those facts already retrieved?' This approach is based upon so-called data induction algorithm. However, in the past decade a lot of research has been done in the field of automated classifiers, sometimes referred to as artificial neural networks. Their approach to information is based upon a totally different concept to modelling, as our PFR2 database; they create the models themselves. Nevertheless, there are a few problems associated with this type of retrieval. First, substantially more processing and storage power is required; secondly, a training of the particular algorithm is required. The first problem will, hopefully, be resolved by time as the technology advances; the second problem is more conceptual. It is quite predictable that in the beginning the induction will cause a tautology retrieval (at least a tautology for human experts), e.g. 'In Africa all plants live in the temperature of 30°C'. However, after this initial 'teaching' the induction might retrieve facts like 'this genus X seems to replace genus Y every time the climate

changed to a cooler zone', etc. This last type of retrieval is closely related to aspects of data presentation and their interpretation

## Data presentation

Essentially, we can distinguish between two forms of presentation, textual and graphical. The textual presentation is in fact a report of textual strings extracted from the current database content. The form of layout is usually specified by a given report type, as generated by a report generator usually built-in a given DBMS or created by an application performing query on the behalf of its user.

The other format for presenting results is graphical presentation. Here, this is a palaeogeographic map with the distribution of occurrences of taxa. However, only locations can be mapped onto a graphical layout whose geographical coordinates are known. While in the textual form a 'non-registered' record could appear with a question mark; such a record will often be skipped in the geographical map layout. Ideally, the scale of geographical coordinates will vary with the scope of the data records. For precise locations a point may be used; for a given locality perhaps a circle with a given diameter and centre location; for a river perhaps a line; for a state, lake, sea, continent, etc. a complex polygon. Those features indicate the requirements on the graphical capabilities linked onto the database structures. These heterogeneous systems are often referred to as a geographical information system (GIS). We have linked the PFR2 database to the palaeomapping software Atlas, a product of the Cambridge Map Company. Figure 2 gives examples of typical maps produced by the whole temporal GIS system, that is PFR2 and Atlas. Maps can be drawn for any time range in the Cenozoic and Mesozoic and for any of the taxa in the database.

Although this type of data presentation is quite intuitive and well-arranged, it has some drawbacks associated with it. The first problem relates to the overlapping of locations. If the scope is large and a particular world domain is rich in given occurrence records, then the location points might merge together and overlap, so at the end it looks like one or two occurrences, while an independent (mostly random or incorrect) occurrence somewhere else may produce similar layout, which then confuses the results. Two possible solutions to this problem have been tried. The first was to use different symbols, which, if overlapping, will produce a picture with minimum of complete overlaps, e.g. circle and triangle instead of squares. This might be improved by colour coding. The second solution is to retrieve and display only those records which fit within a given location statistically; the location is then filled with symbols (of a particular colour) showing that that locality has a substantially important statistical occurrence of that taxon.

The second problem relates to the geographical locations. For example, investigating genera near poles often means that localities, although small in size, appear as visually large areas. This is mainly caused by the projection used to display results and after changing the projection (to a polar one, for instance), such deformation should disappear. Nevertheless, some other locations may be then distorted and the user should be aware of it.

## Some consequences for biology

PFR2 is a widely available electronic database which may contain any amount of data, and all changes and additions and annotations can be signed and timed. Thus the entire spectrum of opinion, both objective and subjective, can be contained within it and reviewed by the users. Thus, the formalities of the International Code of Botanical Nomenclature can be followed and recorded. An individual's recommendation for 'Names in Current Use' status can be recorded (Boulter *et al.* 1991); indeed the Frankfurt Declaration reserved a field specially for this. On the other hand, it becomes increasingly clear that the existence of the database, freely in the public domain, makes many restrictions unnecessary. The temporal features and easy storage of huge amounts of data change the way taxa are conceptualized. In turn, this will change the way we perceive those taxa as biological entities and lineages. Binomial nomenclature that grew up from the days of Linnaeus has survived within the culture of printed information: books, libraries, mail, and more recently photocopies and fax. Within palaeontology the survival has been more difficult than with rating modern species (e.g. for all the reasons making it so impossible to define a fossil species) but it is still used very widely.

## Future developments

We can detect two short-term and one long-term goals. The first two are associated with current PFR2 performance and capabilities, and the latter will lead to a new stage, PFR3.

### (i) Graphic catalogue

The experiments with SuperJANET are to evaluate not only remote search protocols, but mainly the capabilities of transferring images, since the SuperJANET will be a network of an ultra-high bandwidth capable of fast transfer of compressed images. In association with this a fractal compression algorithm is being evaluated. The users would then be able to browse not only text information, but also to view images of fossils, which might improve the data interpretation process. Optionally, CD-ROM technology might be able to substitute SuperJANET at sites which are outside the network. CD-ROM seems to be a good solution, but the updates are expensive and take time.

### (ii) Temporal databases

Generally speaking, every piece of information stored by the database is inherently temporal. The PFR2 database can be viewed as a causal chain of temporal elements stored in the database. This facility would then allow us to create an insight into how a particular domain evolved during time, which may be useful for historical research. It may also uncover patterns in the data registration process, helping to detect unreliable sources, etc. At the current stage of development, PFR2 stores all the temporal information required for a full temporal database; however, the mechanisms and query algorithms have not yet been adjusted.

### (iii) Adaptable system

This last topic is our long-term goal and is still in the stage of preliminary ideas. Nevertheless, let us look at the kernel of a new system which might perhaps lead to PFR3. In order to do that, we have to review the concept of a general database system. It was mentioned above that a database is ideally reflecting a scientific system commonly accepted by its developers and users. What would happen if we put a human expert in the place of the electronic database and investigate what happens then? Even if the expert has the time, power, precision and capability of doing all the tasks the database is doing, there would be one significant difference:

the human expert would be (and probably always is) influenced by the data they are working with, while the electronic database calculates the records in a calm manner as if it were numbers, which in fact they are. This analogy leads us to a new idea of allowing the database to be influenced by its data and perhaps even to be able to 'forget' certain information. The most commonly used data are probably the most reliable and, therefore, should attract the attention of the database. The database should learn those most commonly used patterns and try to fit them within other rarely used data. The data which do not fit within a suggested pattern are either exceptional or unreliable and so they could be stored on a peripheral part of the database freeing the active resources of the adaptive database.

The authors would like to thank our colleagues P. Woolliams, N. Ambihaipahan, S. Brown, D. Gee and D. Woodhouse for stimulating discussions. The work is partly financed by the H.E. Funding Council.

# References

ANDREWS, H. N. 1970. Index of generic names of fossil plants, 1820–1965. *Bulletin of the United States Geological Survey*, **1300**, 1–354.

BLAZER, A. M. 1975. Index of generic names of fossil plants, 1966–1973. *Bulletin of the United States Geological Survey*, **1396**, 1–54.

BOULTER, M. C. & FISHER, H. C. (eds) 1994. *Cenozoic Plants and Climates of the Arctic*. Springer-Verlag, Heidelberg.

——, CHALONER, W. G. & HOLMES, P. L. 1991. The IOP Plant Fossil Record: are fossil plants a special case? *In*: HAWKSWORTH, D. L. (ed.) *Improving the Stability of Names: Needs and Options*. Regnum Vegetabile 123, Koeltz Scientific Books, Konigstein, 231–242.

BOULTER, M. C., BROWN, S. M., KVACEK, J. & LHOTAK, M. 1993. *A Provisional List of Plant Taxa which have Fossil Species*. International Organisation of Palaeobotany, London, Circular **11**.

HARLAND, W. B. *et al.* 1990. A Geological Time Scale 1989. Cambridge University Press, Cambridge.

HOLLIS, S. & BRUMMETT, R. K. 1992. *World Geographical Scheme for Recording Plant Distributions*. Hunt Institute for Botanical Documentation, Carnegie Mellon University, Pittsburgh.

HOLMES, P. L & HEMSLEY, A. P. 1990. *The Revised Format and Content of Plant Fossil Records*. International Organisation of Palaeobotany, London, Circular **8a**.

HUGHES, N. F. 1989. *Fossils as Information*. Cambridge University Press, Cambridge.

MEYEN, S. V. 1991. Catalogue of Fossil Plants. International Organisation of Palaeobotany, London, Circular **9**.

RAVN, R. 1989. *Taxon*. Privately produced computer database.

TRALAU, H. 1974. *Bibliography and Index to Palaeobotany and Palynology, 1950–1970*. Swedish Museum of Natural History, Stockholm, **261** and **358**.

——& LUNDBLAD, B. 1983. *Bibliography and Index to Palaeobotany and Palynology, 1971–1975*. Swedish Museum of Natural History, Stockholm, **206** and **239**.

WATT, A. D. 1982. Index of Generic Names of Fossil Plants, 1974–1978. *Bulletin of the United States Geological Survey*, **1517**, 1–63.

ZADEH, L. A. 1972. A fuzzy set theoretic interpretation of linguistic hedges. *J. Cybernetics*, **2**, 4–34.

*Note added in proof.* The latest version of the database, PFR 2.2, is on the Internet at: http://www.ucl.ac.uk./palaeo/.

# Structuring soil and rock descriptions for storage in geotechnical databases

D. G. TOLL & A. J. OLIVER

*School of Engineering, University of Durham, Durham DH1 3LE, UK*

**Abstract:** The paper describes the development of a database for geotechnical information. The database is part of a knowledge-based (expert) system for the interpretation of geotechnical information. It has been implemented using the INGRES relational database management system. A particular feature, unique to the database structure developed, is the ability to store layer descriptions in a structured form. In other database systems these are simply stored as single text fields. A complex structure has been derived which can accommodate the full range of layer descriptions; for example, interbedded soils/rocks within a layer, the variation in description from top to bottom of a layer, and the existence of multiple soil or rock types within a description. The paper presents details of the database tables with examples of how the structured representation is used.

This paper describes the development of a database for geotechnical information. It has been developed as part of a knowledge-based (expert) system (called SIGMA) for the interpretation of geotechnical information (Toll *et al.* 1992; Toll 1994). The role of the knowledge-based system (KBS) is to assist the geotechnical specialist with both the interpretation of ground conditions across a site and the assessment of design parameters from numerical information (laboratory and field test results). The database forms the core of the system, and holds the data on which any interpretation is based.

A particular feature, unique to the database structure developed, is the ability to store layer descriptions in a structured form. In other database systems these are simply stored as single text fields. The text string would have to be parsed each time in order to abstract information from the description. Parsing (in its widest sense) is the process of breaking down a sentence or phrase into its constituent parts and recognizing their relationship with other words from the same phrase or sentence. This requires an understanding of the grammer and syntax of the domain being examined. The grammer and syntax of soil descriptions is not as complex as natural language, nevertheless it still has a rich vocabulary. Attempting to abstract information from text strings using standard database search routines (based on single key words) will not work for complex descriptions since such an approach cannot incorporate any recognition of syntax. For instance, in the description 'strong SAND-STONE with bands of weak LIMESTONE', a simple search facility could pick out the words 'strong' and 'weak' but would not be able to link those terms with the appropriate rock types.

SIGMA needs to make use of the layer descriptions in order to aid the interpretation process; both for interpolating between boreholes and for checking consistency between field descriptions and test results. Parsing the description each time information is required would be very inefficient. To parse one record of 30 terms containing two strata, each having several constituents, takes less than 5 s within the development environment of SIGMA. However, to parse all the descriptions for an investigation involving a large number of boreholes, each containing a number of layers, would significantly degrade the response time of an interpretation system. SIGMA has been developed as an interactive tool, and such delays would cause frustration to the users.

Therefore, the layer description is parsed to as low a level as possible at the data entry stage, and these parsed data are stored. This has led to a complex structure to accommodate the full range of layer descriptions; for example, interbedded soils/rocks within a layer, the variation in description from top to bottom of a layer, and the existence of multiple soil or rock types within a description.

The KBS has been developed using the ProKappa expert system development environment (Intellicorp 1991). The database has been implemented as a stand-alone system using relational database management system INGRES (Relational Technology 1990). Communication between the database and the KBS is via ProKappa Data Access routines utilizing SQL (Structured Query Language). Access to the database by other packages is possible using the SQL standard and ASCII file transfer.

Although the structure may appear complex, that will not be apparent to users of SIGMA. The

*From* Giles, J. R. A. (ed.) 1995, *Geological Data Management*,
Geological Society Special Publication No 97, pp. 65–71.

user interface which forms part of SIGMA ensures that users never have to make up the complex SQL required to directly access data themselves. The SQL statements are constructed by SIGMA from the responses the user has made to specific questions. The system knows what data reside in which tables and is able to abstract data from various tables and combine them if required.

## An overview of the database

An outline schema for the database is shown in Fig. 1, where the boxes represent tables in a relational database structure. In developing the schema, use has been made of the Association of Geotechnical Specialists' format (AGS 1992, 1994). This format was evolving during the period of the database development and has since been adopted as an industry standard for exchange of site investigation data.

Each table stores data which represent a data group, the data group being a function, property or parameter of the Site Investigation. Tables are linked together via keyed fields, that is fields that are common to both tables, and allow data to be aggregated over many tables. This structure, derived through third-order normalization techniques, produces an efficient structure for data retrieval and handling, necessary for the potentially large volume of data to be stored. Normalization is the process whereby conceptual data models are transferred into a form acceptable to a relational database (Codd 1970). The result of the normalization process is a data model that has a minimum level of duplication and redundancy, the relationships between the attributes, fields, are clearly shown and a more flexible data model is produced (Bamford & Curran 1987; Date 1987).

As far as possible the table names have been adopted to be compatible with the AGS headings. The top level table is the project (*proj*) table which contains information on the location and date of the project and the parties involved. A geology (*geol*) table has been linked to the *proj* table through the *proj_id* key. The *geol* table allows storage of identified geological horizons which could exist at the site. This stratigraphic information will generally be obtained from a desk study at the feasibility stage of the project before any ground investigation has been started (boreholes or trial pits). Therefore the information need not be related to specific holes but is attached at the project level. The information can be linked to specific layers identified at a later stage (during or after the ground investigation) using the *horz_no* field.

The term 'hole' has been adopted as the generic name for boreholes, trial pits or shafts (as per AGS). The *hole* table contains details of boreholes or trial pits such as location, date and method of boring. The details of the ground conditions observed at the hole are stored in the layer (*lay*) table. This contains depth and thickness information about the layers observed, and these can be linked to the appropriate geological horizon in the *geol* table. Also present is a text field containing the soil or rock description. The reasons for maintaining this text field are described below.

Minor comments on the ground conditions which are identified by a particular depth, and which do not correspond with layer depths, are not stored in a structured form but are held as text fields in the depth related remarks (*drem*) table, attached at *hole* level. Fracture spacing data are also stored separately in the *frac* table. This is also attached at the *hole* level, rather than being identified with a particular layer (Fig. 1). This is because zones of similar fracture spacing identified will not necessarily coincide with layer boundaries.

Test results from laboratory and *in situ* tests are stored within individual tables for each test. The database structure allows multilevel storage of test data; the more frequently accessed data are available separately from the raw or semi-processed data. This aspect of the database structure is described by Oliver & Toll (1995).

## Layer descriptions

Engineering descriptions of layers used on borehole and trial pit logs can be long and complex; for example, 'Moist reddish brown stiff thinly bedded closely fissured silty sandy CLAY with a little dark greenish grey sub-rounded fine gravel and frequent inclusions of sand'. If information needs to be abstracted from the description, say the consistency of the soil (*stiff*), the text description must be parsed. Rather than parse a description each time a piece of information is needed from it, a more efficient practice is for the description to be parsed once, and once only, at the time of data entry. The information contained in the description would then be stored as separate fields within the database, and would be easily and efficiently accessible.

To develop data structures for storing the information contained in soil and rock descriptions is not straightforward. A large amount of varied information is contained in the description and the vocabulary used is often complex.

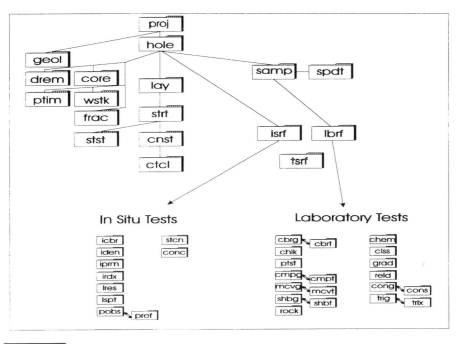

**Fig. 1.** Schema for the SI database structure.

| LEGEND | | | | |
|---|---|---|---|---|
| proj | Project | drem | Depth Related Remarks |
| hole | Borehole / Trial Pit | core | Core |
| geol | Geology | frac | Fracture Detail |
| lay | Layer | ptim | Progress with Time |
| strt | Strata | wstk | Water Strike |
| stst | Stratum Structure | samp | Sample |
| cnst | Constituent | spdt | Specimen Detail |
| ctcl | Constituent Colour | isrf | In Situ Test Index |
| tsrf | Test Reference | lbrf | Laboratory Test Index |

| LEGEND - Laboratory Tests | | | | |
|---|---|---|---|---|
| cbrg | California Bearing Ratio - General | cbrt | California Bearing Ratio - Detail |
| cmpg | Compaction Tests - General | cmpt | Compaction Tests - Detail |
| mcvg | Moisture Condition Value - General | mcvt | Moisture Condition Value - Detail |
| shbg | Shear Box Testing - General | shbt | Shear Box Testing - Detail |
| cong | Consolidation Tests - General | cons | Consolidation Tests - Detail |
| trig | Triaxial Tests - General | trix | Triaxial Tests - Detail |
| clss | Classification Tests | grad | Particle Size Distribution |
| reld | Relative Density | chem | Chemical Properties |
| chlk | Chalk Crushing Value | ptst | Laboratory Permeability Tests |
| rock | Rock Properties | | |

| LEGEND - In Situ Tests | | | | |
|---|---|---|---|---|
| icbr | In Situ California Bearing Ratio | iden | In Situ Density |
| iprm | In Situ Permeability | irdx | In Situ Redox |
| ires | In Situ Resistivity | ispt | Standard Penetration Test |
| stcn | Static Cone Penetration | conc | Cone Calibration |
| pobs | Piezometer | pref | Piezometer - Detail |

Therefore the task of developing a representation scheme which can handle the full range of information is difficult. While much of the very detailed information contained in the description plays only a minor role in engineering design, there may be occasions when it is required. A representation scheme which can handle the majority of soil descriptions was put forward by Toll *et al.* (1991), and this scheme has been extended during the implementation of the database.

Since it is possible that the structured representation will not be able to handle all the esoteric descriptions that can be found on borehole logs, provision has been made in the layer (*lay*) table to store the description in its full form as a text field. Therefore, the full description is always available to the engineer or geologist processing the data. However, interpretation of the data by the KBS will use the structured representation. It will be seen that the structured representation is still very detailed; it might even be argued that it is too detailed for the purposes of most investigations. However, it is felt that a reasonable compromise has been reached on the ability to represent very complex descriptions without carrying too much redundant information.

In order to represent the complexity of soil and rock descriptions four tables are required: *strt*, stratum; *stst*, stratum structure; *cnst*, constituent; and *stcl*, colour. A listing of these tables in the database, their data fields, data types and a description of each field's purpose is given in Appendix 1.

The stratum (*strt*) table is required because descriptions of layers may contain more than one stratum (soil or rock). For example, 'SAND-STONE interbedded with SILTSTONE' or 'CLAY with pockets of SAND'. In these examples two distinct strata are present within the layer, yet they cannot be distinguished as separate layers (a layer being defined by depth and thickness). Therefore the representation scheme allows for the possibility of multiple strata within a layer (identified by stratum number, *strt_no*). The relationships between the strata are stored in the stratum structure (*stst*) table. In the first example given above, the term 'interbedded' would describe both the strata sandstone and siltstone. In the second example the term 'pockets' would describe the sand while clay would be described as 'dominant'.

This format can also deal with the case where a soil or rock changes significantly from the top to the base of a layer, e.g. 'silty sand becoming clayey sand'. The first stratum, silty sand, would

be identified in the stratum structure (*stst*) table as 'Top', and clayey sand as a separate stratum identified as 'Base'. Other information on the structure can be recorded such as bedding dip, orientation and spacing, or spacing of inclusions.

'Stratum' represents the whole stratum, e.g. silty sandy clay, whereas 'constituent' indicates the constituents combining to make up the stratum, e.g. silt, sand, clay, etc. In the stratum (*strt*) table, information which relates to the whole stratum is stored such as moisture condition, consistency, plasticity or weathering, and the dominant constituent is also included. In the constituent (*cnst*) table, information which relates to individual constituents is stored. The amount of each constituent is identified as 'main' if the constituent is dominant (silt, etc.), or as 'minor' (slightly silty, etc.), secondary (silty, etc.), or 'major' (very silty, etc.) for the lesser constituents. This scheme can also be used to represent descriptions which are not those recommended by British Standard 5930 (1981), but which are still in use, such as 'with some ...', 'with a little ...', etc.

Since it is sometimes possible for detailed information on grading, shape and texture to relate to a particular constituent, rather than the stratum as a whole (e.g. 'clay with a little black subrounded coarse gravel', where subrounded and coarse refer to the lesser constituent, gravel) this information is stored in the constituent table, rather than at the stratum level. Colour can also relate to the individual constituents, rather than to the stratum as a whole (as when the lesser constituent gravel is described as black in the example above). Colour is represented as main colour, secondary colour and modifier. Since there can be multiple colour descriptors for a constituent a separate colour (*stcl*) table is used to handle this level of complexity, attached at constituent level.

The description of weathering is difficult and different terminology is used depending on the material being described. While the database field can be used to store any descriptive term for weathering, it may be necessary to restrict the terms which the KBS can recognize, and therefore use in the interpretation process. Further work is required on this aspect.

Examples of the way in which the representation scheme would be used are shown below for three layer descriptions.

- *Layer 1.* Moist reddish brown stiff thinly bedded closely fissured silty sandy CLAY with a little dark greenish-grey rounded fine gravel and frequent inclusions of sand.

- *Layer 2.* Thinly spaced layers of moist dark yellowish-brown firm CLAY interbedded with thickly spaced firm blue-grey SILT. Bedding dip 167/05°.
- *Layer 3.* Brownish grey becoming brown thickly bedded slightly weathered oolitic LIMESTONE. Strong.

## Parsing

To enter layer descriptions into the structured database format would be tedious, so an automated parser has been developed by Oliver (1995), based on work by Vaptismas (1993). It allows the layer description to be broken down into its constituent parts and stored for use in the interpretation process. The parser comprises numerous functions or clauses, each clause being responsible for the identification of particular geotechnical terms or phrases. The grammar and syntax of a layer description adopted is based on British Standard 5930 (1981), but provision has also been made for other non-standard forms which are currently in use in the geotechnical community. The parser must be capable of identifying not only the word or

**Table 1.** *Stratum (strt) table*

| proj_id | hole_id | lay_no | strt_no | strt_mstp | strt_mscd | strt_cons | strt_psty | strt_wthg |
|---------|---------|--------|---------|-----------|-----------|-----------|-----------|-----------|
| 7702 | B5 | 1 | 1 | CLAY | MOIST | STIFF | | |
| 7702 | B5 | 1 | 2 | SAND | | | | |
| 7702 | B5 | 2 | 1 | CLAY | MOIST | FIRM | | |
| 7702 | B5 | 2 | 2 | SILT | | FIRM | | |
| 7702 | B5 | 3 | 1 | LIMESTONE | | STRONG | | SLIGHTLY |

**Table 2.** *Stratum Structure (stst) table*

| proj_id | hole_id | lay_no | strt_no | stst_no | stst_stct | stst_spac | stst_dip | stst_ornt | stst_surf | stst_dcon |
|---------|---------|--------|---------|---------|-----------|-----------|----------|-----------|-----------|-----------|
| 7702 | B5 | 1 | 1 | 1 | DOMINANT | | | | | |
| 7702 | B5 | 1 | 1 | 2 | BEDDING | THIN | | | | |
| 7702 | B5 | 1 | 1 | 3 | FISSURE | CLOSE | | | | |
| 7702 | B5 | 1 | 2 | 1 | INCLUSION | FREQUENT | | | | |
| 7702 | B5 | 2 | 1 | 1 | INTERBEDDED | THIN | 5 | 167 | | |
| 7702 | B5 | 2 | 2 | 1 | INTERBEDDED | THICK | 5 | 167 | | |
| 7702 | B5 | 3 | 1 | 1 | BEDDING | THICK | | | | |

**Table 3.** *Constituent (cnst) table*

| proj_id | hole_id | lay_no | strt_no | cnst_no | cnst_type | cnst_amnt | cnst_grdg | cnst_shp | cnst_txt | cnst_dstb |
|---------|---------|--------|---------|---------|-----------|-----------|-----------|----------|----------|-----------|
| 7702 | B5 | 1 | 1 | 1 | CLAY | MAIN | | | | |
| 7702 | B5 | 1 | 1 | 2 | SILT | SECONDARY | | | | |
| 7702 | B5 | 1 | 1 | 3 | SAND | SECONDARY | | | | |
| 7702 | B5 | 1 | 1 | 4 | GRAVEL | MINOR | FINE | ROUNDED | | |
| 7702 | B5 | 2 | 1 | 1 | CLAY | MAIN | | | | |
| 7702 | B5 | 2 | 2 | 1 | SILT | MAIN | | | | |
| 7702 | B5 | 3 | 1 | 1 | LIMESTONE | MAIN | | | OOLITIC | |

**Table 4.** *Colour (stcl) table*

| proj_id | hole_id | lay_no | strt_no | cnst_no | stcl_no | stcl_mcl | stcl_scl | stcl_mod | stcl_strc |
|---------|---------|--------|---------|---------|---------|----------|----------|----------|-----------|
| 7702 | B5 | 1 | 1 | 1 | 1 | BROWN | RED | | |
| 7702 | B5 | 1 | 1 | 4 | 1 | GREY | GREEN | DARK | |
| 7702 | B5 | 2 | 1 | 1 | 1 | BROWN | YELLOW | DARK | |
| 7702 | B5 | 2 | 2 | 1 | 1 | BLUE | | | |
| 7702 | B5 | 2 | 2 | 1 | 2 | GREY | | | |
| 7702 | B5 | 3 | 1 | 1 | 1 | GREY | BROWN | | TOP |
| 7702 | B5 | 3 | 1 | 1 | 2 | BROWN | | | BASE |

phrase but its context within the description to ensure that the correct meaning is established. Once parsed, the individual elements are sorted into their correct tables and placed into the database.

## Conclusions

A geotechnical database has been developed as part of a knowledge-based system for the interpretation of geotechnical information. The database has been implemented using the relational database management system, INGRES.

In current geotechnical databases, the engineering descriptions of soils and rocks are stored as single text fields. In order to abstract a particular feature from the description the full text string has to be parsed each time. In this paper it has been shown that the descriptive terms can be stored as separate database fields. Database structures have been developed which can store soil and rock descriptions in a structured form. These structures can allow very complex layer descriptions including interbedded soils/rocks within a layer, a variation in description from top to bottom of a layer and the existence of multiple soil or rock types within a description.

Since the layer descriptions can be represented in a structured form within the database, this means the description only needs to be parsed once. Thereafter, data for a particular aspect of the description can be directly extracted from the appropriate database field.

## References

AGS 1992. *Electronic Transfer of Geotechnical Data from Ground Investigations*. Publication AGS/1/92, Association of Geotechnical Specialists, Wokingham.

—— 1994. *Electronic Transfer of Geotechnical Data from Ground Investigations*, 2nd ed, Association of Geotechnical Specialists, Camberley, Surrey.

BAMFORD, C. & CURRAN, P. 1987. *Data Structures, Files and Databases*, MacMillan Education Ltd, Basingstoke, Hampshire.

BRITISH STANDARD 5930. 1981. *Code of Practice for Site Investigations*. British Standards Institution, London.

CODD, E. F. 1970. A relational data model for large shared data banks. *Communications of ACM*, **13**, 6, 377–387.

DATE, C. J. 1987. *Database: A Primer*. Addison-Wesley Publishing, Reading, MA.

INTELLICORP 1991. *PROKAPPA Users Guide, Intellicorp PK2.0-UG-2*. Intellicorp Inc., CA.

OLIVER, A. J. 1995. *A Knowledge Based System for the Interpretation of Site Investigation Information*. PhD Thesis, University of Durham.

—— & TOLL, D. G. 1995. *A Computer System for Site Investigation Data Management and Interpretation*. Proc. Int. Conf. Advances in Site Investigation Practice, London, March 1995.

RELATIONAL TECHNOLOGY 1990. *An Introduction to INGRES, HECRC, ISG 245*. Relational Technology Inc, CA.

TOLL, D. G. 1994. *Interpreting Site Investigation Data using a Knowledge Based System*. Proc. 13th Conference of Int. Soc. Soil Mechanics and Foundation Engineering, New Delhi, Vol.4, Oxford, New Delhi, 1437–1440,.

——, MOULA, M., OLIVER, A. & VAPTISMAS, N. 1992. *A Knowledge Based System for Interpreting Site Investigation Information*. Proc. International Conference on Geotechnics and Computers, Paris, Presses de l'École Nationale de Ponts et Chaussées, Paris, 607–614.

——, MOULA, M. & VAPTISMAS, N. 1991. *Representing the Engineering Description of Soils in Knowledge based Systems*. In: RZEVSKI, G. & ADEY, R. A. (eds.), *Applications of A.I. in Engineering VI*, Computational Mechanics Publications, Southampton, 847–856.

VAPTISMAS, N. 1993. *A Methodology for the Interpretation of Ground Conditions from Borehole Information*. PhD Thesis, University of Durham.

## Appendix

*Parsed layer description data tables in GeoTec database*

lay

| Field Name | Field Type | Field Description | Remarks |
|---|---|---|---|
| proj_id | varchar(10) | Project / Site Investigation Code | Key |
| hole_id | varchar(10) | Borehole / Trial Pit Code | Key |
| lay_no | i4 | Layer Number | Key |
| lay_dtop | f4 | Depth to top of layer (m) | |
| lay_thck | f4 | Layer thickness (m) | |
| lay_des | varchar(300) | Description of layer | |
| horz_no | i4 | Horizon number | |

strt

| Field Name | Field Type | Field Description | Remarks |
|---|---|---|---|
| proj_id | varchar(10) | Project / Site Investigation Code | Key |
| hole_id | varchar(10) | Borehole / Trial Pit Code | Key |
| lay_no | i2 | Layer Number | Key |
| strt_no | i2 | Stratum number | |
| strt_mstp | varchar(30) | Main stratum type | |
| strt_mscd | varchar(30) | Moisture condition | |
| strt_cons | varchar(30) | Consistency | |
| strt_psty | varchar(30) | lasticity | |
| strt_wthg | varchar(30) | Weathering | |

stst

| Field Name | Field Type | Field Description | Remarks |
|---|---|---|---|
| proj_id | varchar(10) | Project / Site Investigation Code | Key |
| hole_id | varchar(10) | Borehole / Trial Pit Code | Key |
| lay_no | i2 | Layer Number | Key |
| strt_no | i2 | Stratum Number | |
| stst_no | i2 | Structure number | |
| stst_stct | varchar(30) | Layer Structure feature | |
| stst_spac | varchar(30) | Layer spacing | |
| stst_dip | varchar(30) | Dip | |
| stst_ornt | varchar(30) | Orientation | |
| stst_surf | varchar(30) | Surface | |
| stst_dcon | varchar(50) | Discontinuity modifier | |

cnst

| Field Name | Field Type | Field Description | Remarks |
|---|---|---|---|
| proj_id | varchar(10) | Project / Site Investigation Code | Key |
| hole_id | varchar(10) | Borehole / Trial Pit Code | Key |
| lay_no | i2 | Layer Number | Key |
| strt_no | i2 | Stratum number | Key |
| cnst_no | i2 | Constituent number | |
| cnst_type | varchar(20) | Constituent Type | |
| cnst_amnt | varchar(20) | Constituent Amount | |
| cnst_grdg | varchar(30) | Grading | |
| cnst_shp | varchar(30) | Shape | |
| cnst_txt | varchar(30) | Texture | |
| cnst_dstb | varchar(30) | Distribution | |

stcl

| Field Name | Field Type | Field Description | Remarks |
|---|---|---|---|
| proj_id | varchar(10) | Project / Site Investigation Code | Key |
| hole_id | varchar(10) | Borehole / Trial Pit Code | Key |
| lay_no | i2 | Layer Number | Key |
| strt_no | i2 | Stratum number | Key |
| cnst_no | i2 | Constituent number | Key |
| stcl_clno | i2 | Colour number | |
| stcl_mcl | varchar(30) | Main colour | |
| stcl_scl | varchar(30) | Secondary colour | |
| stcl_mod | varchar(30) | Colour Modifier | |
| stcl_strc | varchar(30) | Colour Structure | |

# The use of text databases in the management of exploration data

S. C. R. MALLENDER

*Mallender Hall Ltd, 12 Glenfield Road, London W13 9JZ, UK*

**Abstract:** The management of exploration data concerns all the things that happen to a piece of exploration data during its life, and the management of the physical data is controlled by the information used to describe it. A description of the data forms part of a contract between a company and a recording contractor before the physical data even come into existence. Descriptive information about each data item and each dataset is recorded and re-recorded on many transmittals as the physical data move from recording contractor, to processing contractor, to storage contractor. The 'computerized catalogue' listing, which a storage contractor provides to an exploration company is often the only record of what is available, and the legal ownership records of these data are often stored elsewhere. The amount of this reference information used to describe exploration data (for example, the name and extent of a map or a line, the title of a report, the reference number of a tape) varies significantly from one data type to another, and the amount actually recorded often varies from one record to the next. Historically, there has been a lack of consistency throughout the industry in the terms used, and the number of those terms employed when describing and referring to exploration data. This poses problems in the management of exploration data.

The main problem concerns the storage, retrieval and transmission of this descriptive information; the information is rarely 'processed' as such. From a database point of view, the problem is to store large quantities of quite disparate and unstructured information, to find rapidly what has been stored, and to retrieve the information easily in a (re-)usable form.

Text databases are ideal in these circumstances: they can hold completely unstructured information, and allow users to find records using any data that appear anywhere in the content of the record. Accompanying software is used to provide a user interface to the text database, and this enables sophisticated systems which are very user-friendly to be developed easily.

The first half of this paper describes the facilities provided by a text database and accompanying software which allow useful systems to be developed. It describes how the database supports unstructured data, and how the index, which supports the retrieval of those data, works.

The second half describes how the features of the resulting system can be exploited to deliver significant benefits both to the administrator who manages the data, and to the geoscientists who retrieve and use the data.

The paper is not based on any particular software product, or system that has been developed, though there are a number of both products and systems in the industry.

## The text database

### Record structure

Record structures for any database comprise a number of fields. Each field is used to hold one particular element of data from the record. If the data element is numeric, the field is of type numeric. If the data element is character, the field is of type character. Each field can be described by its type (character, numeric, date and so on) and its length.

In text databases, record structures also comprise a number of fields, and the basic field types (character, numeric, date) are also present. However, text databases are particular in three important aspects.

- In addition to the basic field types there is a field type called text, and most of the fields in a text database record are of this type.
- No field lengths are defined for any field. In practice this means that any amount of data, such as the entire textual content of a special publication, can be stored in one field of one record.
- Any one field can have any number of occurrences within a single record. This means that the whole collection of special publications could be stored in one field of one record, each publication in its own occurrence of the field. A more practical example might be a record which holds a CV, with a field called 'workExperience', which holds each paragraph of work experience as a separate occurrence within the field.

*From* Giles, J. R. A. (ed.) 1995, *Geological Data Management,*
Geological Society Special Publication No 97, pp. 73–79.

The most appropriate way to represent a text database record pictorially is in three dimensions. Figure 1 illustrates the difference between a 'conventional' record and a text database record.

A number of features are inherent from the text database record structure. The user has control over every record, every field in every record, and every occurrence of every field. Hence when printing out a CV, the geoscientist can print out his or her work experience in chronological or reverse chronological order, or only those paragraphs which relate to international assignments.

Occurrences of fields can be linked together. For example, the occurrences of the field duration, employer and workExperience, could be linked. This could be used to present the work experience paragraphs with employer name in order of the longest assignment first, or a summary table of the employer of each assignment and its duration. Note that this kind of linkage relationship exists within a record, not between records.

There is another kind of relationship that exists within this record: it is the one-to-many

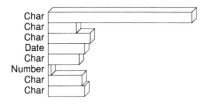

## A single 'conventional' record

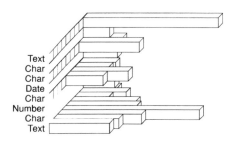

## A single 'text' record

**Fig. 1.** The difference between a 'flat' conventional record, and the 'three-dimensional' text record: in the text record, each field can have any number of occurrences of any length.

relationship between the one surname and the many linked workExperience occurrences.

So far the presentation of a single record has been discussed; it is of course possible to present any number of CV records sorted by any attribute within the CV, and still have control over the internal presentation of each CV. As with 'conventional' databases, it is possible to link together two different databases using a common field. If another database contained the addresses of the employers, the two databases could be 'joined' on the common employer field, to 'pull in' the employer addresses when printing the CVs.

In addition to supporting complex relationships, a text database is extremely efficient on disc space. As the text database does not know in advance how many data will be stored in one field, or how many occurrences of that field there will be, the disc space required for its data files is not allocated in advance, but used as data arrives. A text database field is only as long as the data it contains: there are no blank characters between the end of the data and the end of the field, and fields which are empty occupy no space on disc. The result is that the database files of a text database are full of data, and hence extremely efficient on disc space and data transfer. Text databases can also be used to store and retrieve binary large objects (such as image files) very efficiently.

Text databases can store any type of data, any amount of data, and support complex relationships within the data; however, without sophisticated control, these data could easily become impractical to manage. It is the index which allows the user to find and retrieve data in a useful way.

### The index

The index of a book is an ordered list of terms which refers back to concepts in the main text. The index of a database is an ordered list of terms which refers back to records in the main database.

In conventional structured databases, one field of each record is selected as the 'key' field, and it is the content of this field which is indexed, and produces all the terms in the index list.

In an unstructured text database the concept of the 'key' field is extended to all fields; every field can be indexed, and indexed in different ways. For the fields which are of type character or text, each word in the text has its own entry in the index. For all the other field types except text, the content of each field, and each occurrence of each field also has its own entry in the index.

In the CV example, every word in the workExperience text would be indexed, as would be each individual word of the employer name, as well as the employer name as a whole.

Figure 2 shows the relationship between the data file, and the corresponding index file.

In the ordered list of terms in the index, each term only has one entry. This entry stores the number of times the term appears, and stores vectors which point to where the terms appear in the data. For each point where a term appears, these vectors store the record, the field, the occurrence of the field, and at which point in a particular paragraph or sentence the term appears. (They are referred to as 'vectors' because they indicate not just which record, but how far along the record the terms are.)

Figure 3 presents the index file as a long vertical list, and highlights the presence of one index entry.

Hence, all a user needs to find any record, is a term which appears anywhere in that record. Once supplied and matched to the index, the text database reports how many times the term occurs, in how many records it occurs and can show exactly where in each of those records the term appears. Further, a well-designed system will supply these results with sub-second response times for databases in excess of a

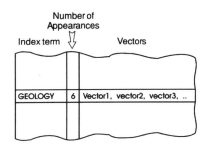

**Fig. 3.** An example of one record in the index file, showing that actual term, the number of occurrences and the vectors.

million records. There are two reasons which account for this fact.

- As data entries are added to a text database, the index list grows in length. However the number of new terms, those which generate new entries in the index, rapidly diminish. For a given type of data, most terms have been extracted into the index by the time the first 25 000 records have been added. The next 1 000 000 records hardly affect the length of the index at all.
- In addition, each term is 'hashed' before being stored in the index, i.e. an algorithm is used to produce a binary string representation of the index term. When a term is supplied by the user to be matched against the index, this term is also 'hashed' by the same algorithm and comparison takes place in a bit-wise fashion. This means that comparison does not start at the top of the index list and work down until a match is found; instead it enables the correct index entry to be located directly, and the entire index entry record is read into memory within about three disc reads. (Disc reads take a few milliseconds.)

In practice a user rarely uses a single criterion to specify the required records. Criteria are combined using Boolean logic (for example, '(international OR CIS) AND Fina') and searches are restricted to specific extents or date ranges. In this case, each single criterion has first to be found before the index records can be compared; however, the presence of logical AND's rapidly reduces the resultant sets.

The text database index provides three features which may not immediately be apparent:

- an index list can be displayed for each (non-text) field, for example, a list of all employers could be displayed; terms can be selected from this list to prepare a search statement;

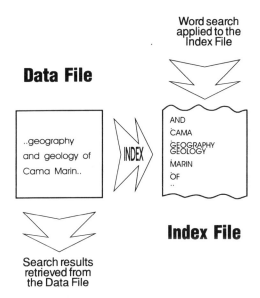

**Fig. 2.** The full text in the data field is indexed to produce the list of terms in the index file. The word search is applied to the index file, which has pointers back to the records in the data file, from which the search results are retrieved.

- an index list can be displayed for any term; this presents a vocabulary list which starts or ends with a particular stem, for example, biostrat*, and again terms may be selected from the list to prepare a search statement;
- a search can be applied to the index files of any number of quite different databases simultaneously, as all the index files have the same structure. The text database handles the complexities and the user appears to be searching a single database, whereas in fact, there may be many databases containing different types of data.

## Developing a useful system

Systems typically comprise a database, some processing, and some screens which are the user interface. Text database systems are no different and a typical text database system comprises:

- databases to hold the data;
- entry screens to get the data in;
- search screens to find the data again;
- output screens to present the data found;
- processing to enhance the system.

Figure 4 schematically illustrates these elements.

To complete the application, some navigational elements such as menus or function keys are provided to allow the user to move around between the screens.

The software which accompanies the text database is fourth-generation software, i.e. the users' system is not constructed by writing programs, but rather the designer selects options and completes boxes on the design screens, and the fourth-generation language software 'constructs' the users' system. This means that all elements of the system are user-definable, and every screen in the system can be laid out just as users specify. Once designed, the system can be rapidly developed as construction is screen-based rather than program based. Data can then be loaded using batch loading or the entry screens, and the system is ready for use.

## Summary

Before considering how text databases can be exploited in the management of exploration data, a summary of the relevant features is presented:

- text databases can support any type of data, any amount of data and extremely complex relationships within the record structures;
- text databases allow a user to find any record by any one term, or any combination of terms, within it, and can display index lists for any field, or any stem;
- text database software is used to develop useful systems, which can be custom-built to users' specification.

## Exploration data management

Before considering how text databases can be exploited to deliver benefits in exploration data management, consider first the requirements of a data management system.

- The data administrator needs to be able to capture structured and unstructured information about exploration data and record it rapidly and accurately.
- In order to find relevant data, the searching user requires a system which is easy to use, and requires no prior knowledge of how or what information has been captured for any particular item of exploration data.
- The results of any search session must be presented in a useable form, such as a list sorted by particular attributes, or a request for a copy of the data item sent directly to the data administrator.

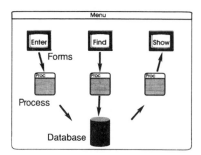

**Fig. 4.** A schematic representation of a text database application, showing the entry form, the search form, the output form and the database; where processing can be introduced between the forms and the database to enhance functionality; and the menu system which ties it all together.

There are other activities involved in the management of data, such as loans for internal items, and orders to off-site storage contractors for items held there, which could be usefully supported by a data management system.

## Exploiting the record structure

To illustrate how a text record supports the attributes required to describe exploration data, consider what type and length of fields are required to hold the data which describe a map.

- A map title can be any length and include any amount of geographic and stratigraphic information (or conversely these can be separated out into other fields).
- Other text fields are required for Type, Author, Contractor, Licence, Round, and Bar code. Where this information is not appropriate for a particular map it may be omitted.
- A date field is also required.

For a text database, the fact that no lengths are assigned to any fields allows as little or as much information as is available to be captured. In addition, it is not important in which fields the information is recorded, as a text search will find information regardless of which field, or where within the field, the information is recorded. Hence, it does not matter if information has been inconsistently recorded; as long as it has been recorded using a text database, the data item can be found.

Consider now seismic information. A seismic line may pass through a number of blocks, and it can be useful to capture this information. This allows a searching user to ask which lines pass through any specific block, or which blocks any specific line passes through. As text databases allow any number of occurrences of data in any one field, many block references can be entered into the one field.

This principle extends to many other types of data; for example, multiple shot-point ranges, log runs, geographic regions, authors and so on.

The one-to-many relationships which can be supported by the text database record structures are useful for holding well information. For each one wellName, there can be many occurrences of the fields associated with the wellLog. This allows a user to find all the logs available for a particular well, or all the wells, for which a particular depth-range, and/or type of, log is available.

Any information which is actually made up of large amounts of real text, such as well reports, for example, are totally at home in a text database. Any amount of information can be captured, from the title to the full textual content of the report. A happy medium might be to capture the abstract, which is likely to contain the key vocabulary which indicates to a user whether or not a report will be relevant to a particular project.

## Exploiting the index

It is the index which provides power to the user to find information easily and rapidly. A user does not need to know how the information was captured, or which attributes have been recorded. A user can call up a list of vocabulary around any stem, or for any field to see what information has been captured there.

For example, a user does not need to know which combinations of the words 'syn.', 'synth.' or 'synthetic', and 'seis', 'seismo' or 'seismogram' have been used to describe a synthetic seismogram. The user requests a list of all the terms in the index which match the stems 'syn' and 'seis'. The system displays the list and the user selects appropriate terms directly from the list. The system then makes the search, retrieving only those records which contain the selected appropriate terms.

A geographic search works the same way. The user starts by requesting a list of all the terms in a field containing geographic reference information. The system displays a list which shows all the geographic areas available. The user selects from the list presented and the system retrieves all the relevant records. This displaying of the index list can be combined with other search features. For example, the user can ask for a list of areas for only those records which contain a particular company's name; sorting these on the frequency of occurrences can show which company is busy where!

A user may not be concerned about where in a particular field or record an element of information is present. For example, a user looking for gamma ray logs in particular is not concerned how the log type information has been captured. The fact that the term 'GR' appears at all in the following combinations is enough for a text database: CNL-GR-CALI, GR-FDL-CALI, CNL-CALI-GR; its position does not prevent the record being found, or the term from appearing in the index list, when such a list of the log types is requested. In a similar manner, a user is uninterested in whether stratigraphic information has been stored in the title field of a map, or in any other field; it has its own entry in the index, and appears when such an index list is displayed.

If a full report is held in a text database record, the user can find the record using any word or combination of words which appear anywhere within the report. Not only that, but the user can see how many times each word of interest appears (which gives an indication of relative relevance), and can go straight to those paragraphs in the report to read the relevant sections.

One of the most useful features of the text database when applied to exploration data management is its ability to apply a search to the index files of any number of quite different databases simultaneously. If the data management system contained a number of databases holding different data types, such as map, seismic, well and bibliographic data, one search can be applied across all databases simultaneously. To the user, the system can appear as one big database, and provides the results of the search 'find any data for region x', as quickly and easily as if a single database were being searched.

## Exploiting the ease of developing a system

It is clear that a 'computerized catalogue' system holding all exploration data can easily be developed using text database software. A user can request a list of the index terms, create a search and view the results from a simple user interface screen. Access to all search features are available via one screen.

However, the fact that the text database software can be exploited to develop useful systems allows a fully integrated Exploration Data Centre Management System to be developed.

During data entry, data values can be validated, or users can select from appropriate lists of acceptable values, for example, when describing the type of well log. Information can be copied automatically from one record to the next, avoiding the re-keying of information which is common to a number of records, for example, when entering seismic line series. Many records can be edited or corrected using a single command, and information can be added to many records simultaneously.

Sub-systems can be developed to support the activities of the Data Centre. For example, a searching user can use the result of a search to send a request for a copy of the physical item to the data administrator. This request can automatically generate a loan request for data held in-house and an order request directed to the relevant storage contractor for data held offsite. A data administrator could use the results of a search to generate a transmittal for those items

directly.

These sub-systems are developed using small text databases and associated screens, and are developed in the same way as the main system. Bar coding can be used to track and manage the progress of data supplied against the requests. (Text databases can store and find records using bar codes as efficiently as convention databases.) The information generated by the Exploration Data Centre Management System can be used to monitor the performance of the storage contractor and data administrator, and manage invoicing.

Another feature which is native to the text database, is the thesaurus. A geographical or stratigraphic thesaurus can be used by an expert user to expand or refine a search.

## Summary

Text databases allow complex record structures to be created which accommodate real data in their natural form. They impose no limits on the amount of data in a single record or a single field. They support textual unstructured information, as well as conventional structured data. They support the relationships which are inherent in exploration data, such as the relationship between the one well, and its many logs.

A text database stores the information required to describe any type of data, from a geophysical tape, to a synthetic seismogram. They allow as much (or as little) information as is available to be recorded for any data item, and further information can be added to existing fields of existing records at any time in the future.

Text databases are very efficient in their use of disc space and machine resources: all data, whether 'free text', image data, or even bar code, are stored in the most compact form. Text databases provide extremely fast responses: no conventional database can approach the speed of a text database for information retrieval.

The unique indexing allows a user to find any records in a text database using any term which appears in any field of those records. The user can display all the entries in the index for any field, or display all the entries in the index which match a specified stem. The user can select any of the terms listed, and combine them with other terms to produce simple and sophisticated search statements. A user can apply a search across any number of databases simultaneously, as easily as searching one database.

This means that a user does not need to know what terms have been used to describe the data items, nor in which fields, or even on which databases they have been recorded.

Using the fourth-generation language software, text database systems can be developed rapidly and configured to the users' specification. This means that they satisfy users' requirements for data entry, searching, and presentation of the results. The way they work is straightforward and obvious, and hence they are very easy to use.

# The geological data manager: an expanding role to fill a rapidly growing need

D. J. LOWE

*Data and Digital Systems Group, British Geological Survey, Keyworth, Nottingham NG12 5GG, UK*

**Abstract:** The management of geological data—raw, inferred and derived—has been neglected for many years. Where such management was attempted, the staff responsible were commonly non-specialists, and most were under-supported and under-valued. Expanding use of digital systems in the geological sciences and growing acceptance of a need for quality management of data in an increasingly business-orientated and competitive world have necessitated reappraisal of data management needs and the role of the geological data manager. The importance of supporting work to develop data verification and validation procedures and the means to assess and ensure data quality is gaining recognition. Some 20 years after the first major incursion of digital technology into the field of geology, the value of the data already gathered, and thus the importance of geological data management, is only now beginning to be appreciated and accepted. In the near future the geological data manager's role is expected to expand rapidly to address the needs of geological activity.

*The environmental sciences, with large quantities of diffuse data, have perhaps the most pressing requirements for improvements of data management procedures of any scientific discipline.*
(Woodland *et al.* 1973, p. 4)

Data management as a specific function has achieved formal recognition only with the introduction, expansion and wide application of digital technology. Management of data ranged between providing indexed and cross-referenced archives, and the more widespread philosophy of 'Stick it in a drawer, it might be useful one day'. The British Geological Survey (BGS) is one example of an organization with a tradition of zealous data acquisition and hoarding; its data holdings, collected by a multitude of individuals, for different purposes and to a variety of specifications, are wide ranging and in many instances unique. Although BGS staff have attempted to collect and pass on reliable information it is inevitable that data quality varies greatly across the corporate data holding. Thus, assessing, labelling and managing data quality is a preoccupation of data managers as well as those collecting or using data.

Past attempts to address (or ignore) data management and quality management requirements are reflected in current data holdings and the scale of inherited problems is only now becoming apparent. Quality managed digital systems require rigorous control in a world where data holdings represent money in the bank, but are of little real value if they are obscure, inaccessible and of unknown quality.

Rather than being the filing clerk of past practice, the modern data manager must have knowledge of the nature, value and limitations of the data, as well as an understanding of the techniques involved in their indexing, storage, retrieval and improvement. Despite the high levels of skill and motivation required, data management remains unglamorous (e.g. Feineman, 1992), and in terms of financial and logistical support it continues to be the poor relation of the higher profile data gathering and data interpretation tasks, which comprise original research.

## The scope and format of geological data

Within a multi-faceted organization such as the BGS, data exist in three basic forms (Fig. 1):

- *Materials*, including raw and processed samples and derived products such as thin sections of rocks, are readily managed, provided that ample space, suitable 'containers' and proper handling facilities are available. This view becomes less simple if we note, for example, that material from one rock sample can be held in many ways, including its original form, as a powder or core plug for analysis, as a polished slab or

*From* Giles, J. R. A. (ed.) 1995, *Geological Data Management*, Geological Society Special Publication No 97, pp. 81–90.

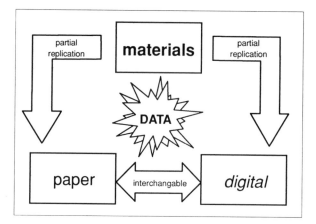

**Fig. 1.** The three basic types of geological data. Data are generally interchangeable between paper and digital systems; information about materials may be held on paper or in digital form.

thin section, and so on. An original sample can also spawn a plethora of 'value-added' information, such as rock or thin-section descriptions, analytical results, fossil determination reports and stratigraphical or geochronometric data.

- *Paper data* include correspondence, manuscript maps, manuscript notes, borehole logs, indexes, catalogues and a variety of published information. The 'value' of these data is commonly historical, even anthropological, as well as scientific. Basic paper data management techniques used in the geological sciences have until recently shown little change since the invention of paper. However, developments of scanning technology and CD-ROM storage/retrieval systems provide tools that facilitate management and manipulation of data previously held on paper, if not of the papers themselves.

- *Digital data*: scanned images, text, relational database tables, graphics files and indexes are some of an increasing number of ways of holding and handling data digitally. Much of this information could be held on paper, but digital systems offer the advantages of improved accessibility, versatility and economy.

Data holdings in each format may be 'raw' (as observed, measured or collected), enhanced, interpreted or derived. Additional information relating to each data type, such as dates of collection, derivation, entry or update, may be linked in the form of metadata. Potential exists for direct or indirect duplication between

material, paper and digital data (Fig. 1). Generally, information that can be held on paper can also be held, manipulated and viewed in digital systems, whether these data be text, graphics or photographs. Duplication of materials in other formats can only be partial and indirect. For example, a fossil specimen can be held as a photograph, hologram or other image, and two- or three-dimensional images can be manipulated using optical or digital methods. Such re-creations cannot, however, fulfil all functions of the material if the original is lost.

More than 20 years ago, reporting on computer applications to data processing, Woodland *et al.* (1973) noted the diffuse nature of environmental science data, but examined neither the detail of this variety nor the many obvious or hidden interrelationships between different data types. Subsequently, rapid advances in computer technology have provided powerful tools that can capture, store and manipulate data, and derive new data efficiently, speedily and cost effectively; but these same digital systems have introduced at least as many problems for the data manager as they have solved.

An unfortunate consequence of digital development is that the more mundane but still essential need to manage non-digital data has attracted relatively little interest or support. The data manager must not only be aware of the interrelationships of material, paper and digital datasets but, within the resources available, must ensure the quality of the data and maintain concurrency between similar information held in different formats.

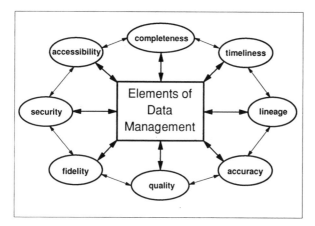

**Fig. 2.** Eight dimensions of data management. The dimensions interrelate in a complex web, but the more significant links are shown by thinner lines around the rim of the figure (modified from Feineman 1992, Fig. 1).

## The question of data quality

Eight dimensions of data management (completeness, accuracy, fidelity, quality, lineage, timeliness, security and accessibility) were discussed and defined by Feineman (1992), who also reported a more general view that data are commonly examined in terms of fitness for purpose and confidence. Complex interrelationships exist between all of Feineman's dimensions of data management, some links being clearer and more fundamental than others (Fig. 2). A subjective division into three groups of dimensions (Fig. 3) provides a view of data

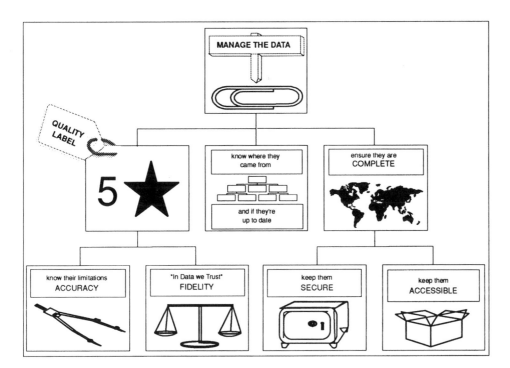

**Fig. 3.** An alternative view of the relationships between data management dimensions. Dimensions of lineage and timeliness in the centre of the figure provide a commonly unseen link between quality dimensions on the one hand and completeness dimensions on the other (dimensional nomenclature after Feineman 1992).

management in which the dimensions of lineage and timeliness form an implicit link between the quality dimensions on the one hand and the completeness dimensions on the other.

From within the eight dimensions, quality stands out as being a composite term whose definition refers back to other dimensions: 'Quality: the data is preferred because of exceptional completeness, accuracy and fidelity' (Feineman 1992, p. 2). Realistically, it is debatable whether completeness need be a criterion of quality. High-quality datasets may be incomplete; while apparently complete datasets may be of inferior quality due to many factors, not the least of which are their lineage and timeliness. The term 'quality' has a variety of implications in different contexts. Problems of assessing data quality vary from organization to organization, and from dataset to dataset, depending on local perception of the term. For example, the BGS holds records of boreholes sunk in the United Kingdom since the earliest days of exploratory drilling. In examining the quality of the data within these records, several fundamental filters may be employed, such as 'Was the borehole cored or was it drilled by open hole techniques?' A second filter is 'Were the cores (if any) examined and described by a reputable geologist, or by someone of unknown ability and qualifications?' Relative values enter the equation, with questions such as, 'Is a description of open hole samples by a geologist more reliable than data from a cored borehole described by an unknown individual?'

In context the ideal information would derive from a complete and well-drilled borehole core examined and described by a geologist. Such information would be seen as fit for most purposes and would be accepted with a high level of confidence. However, even in this ideal case, the record produced and archived could be totally misleading, if depths were mis-measured during drilling, if cores were muddled and mislabelled when placed in their boxes, or if the location and elevation of the borehole site were wrongly surveyed or transcribed.

Even assuming that the data lineage is known and that appropriate operations were completed with accuracy and fidelity to provide a high-quality product, the question of timeliness remains (according to Feineman 1992: 'Timeliness: the data represent the current state of our knowledge'). On reading the record we wonder whether a modern geologist would interpret the cores differently, whether another geologist has amended the original descriptions and on what basis, and whether the original (material or derived) data might still be accessible for checking, even whether terms such as sandstone mean the same now as they did when the core was logged. Borehole records provide a good example of the problem of quantifying quality in a data system, where just a small selection of the potential variables and interactions have been described. Only the original material can be accepted with near total confidence, but even here the potential for human error and subjectivity during collection must be allowed for.

If the geological data manager's remit were to include allocation of subjective Quality Management criteria to all data holdings, then many additional resources would have to be made available; otherwise little time would remain for objective data maintenance tasks. Many data managers will therefore expect to work within an agreed framework of best practice and contribute thoughts and advice on quality issues, but will hope that data collectors and data users will apply any necessary quality labelling. Whether or not such pragmatism succeeds in practice is questionable, probably with different answers in different organizations.

## Verification and validation of data

Needless confusion surrounds the use of the terms 'verification' and 'validation' in the context of data management (Fig. 4).

Data verification relates to processes that establish the basic verity (truth) of the data, however this is assessed. Once this truth is established, the reliability, quality, usefulness or fitness for purpose of the data may be measured, against fixed benchmarks or on a sliding scale, depending upon the uses to which they will be put. Wherever possible, responsibility for verification should lie with those who collect data, and secondarily with those who use it (Fig. 4). Verification can thus be carried out in two stages, the first before data are used, the second during their use, by comparison with adjacent data.

Data validation is potentially a more specific and quantitative process, in which items of information are checked back against original (verified) data or against other data, including lists of valid, legal or legitimate values. Some data management systems include automatic validation checks that compare transcribed data with sets of allowable codes or sensible values (Fig. 4). The 'spellcheck' facility within most popular word-processing packages is a simple example of validation software, which is

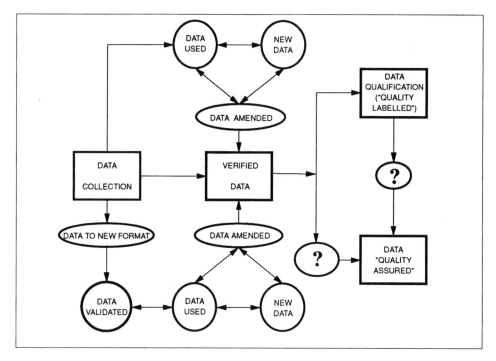

**Fig. 4.** A possible view of the routes through data verification and data validation leading to data qualification. Whether data should be termed quality assured without first being quality labelled in a DQ system remains an open question, as indicated by question marks.

limited by the supporting dictionary and its inability to recognize the use of correctly spelled words in inappropriate contexts. In systems that include conversion of analogue data into digital form, validation routines can be designed to deny entry of invalid information. However, no matter how sophisticated the validation system and regardless of whether it is manual or automatic, it cannot differentiate between data that are correct and those that are incorrect, only between those that are valid or invalid in context. Two examples from the BGS borehole record system illustrate this.

In converting hand-written borehole records to digital format, definitive code dictionaries are used to allow transcription of lithological (rock type) information. 'GRANITE' can be represented by the valid code 'GRAN', but the validation software will not allow entry of a keyboard error such as 'GRAB'. However, 'GRIT', a legal code describing another rock type, would be allowed and pass unremarked by the validation software. Nor can the system recognize that the record of 'GRANITE' in the log might itself be suspect. For instance, a mis-informed driller or quarry manager could describe a hard rock aggregate product, really a greywacke, as 'granite'. This verification rather than validation question should have been noted and rectified by staff who collected the data or requested its transfer to digital form.

The second example involves locational information. Bore site locations are plotted on 1 : 10 000 scale (formerly 1 : 10 560) Ordnance Survey base maps, a practice first used before the National Grid coordinate system was introduced. Some early bore sites, now trans-ferred to modern plans, were derived from sketch maps accompanying the original records, or based on generalized directions, such as '200 yards due west of Home Farm, Thatplace'. When collected, these data were adequate and fit for purpose. Subsequently, first with the introduction of the National Grid (and its adoption as a primary key in borehole indexing) and later with the development of digital modelling techniques, new significance was placed on the quality of locational data for use in more meticulous assessments.

Once a site is marked on a 1 : 10 000 scale map, a National Grid Reference (NGR) 'cor-rect' to 10 m can be measured. Positions appearing still more precise can be measured from the large-scale plans that accompany most modern borehole logs. Although this precision cannot be re-created on the 1 : 10 000 scale site plan, the measured values can be recorded on paper or within a digital index. Nevertheless, the knowledge of an NGR 'correct' to 10, 1 or 0.1 m is no guarantee that the indicated site coincides with that where the borehole was drilled. Automatic systems can check the validity of an NGR within a particular map, but not whether the site was correctly recorded in the first place, precisely remeasured or faithfully transcribed. Nor can such systems judge whether a stated position is the true position, no matter how many digits are quoted in its NGR.

Verification of a borehole's location is a totally different requirement, and commonly an unacceptable position is recognized many years after an original siting, when data are seen to conflict, by not fitting a regular or explicable pattern, during manual or digital surface modelling. The borehole sited '200 yards due west of Home Farm, Thatplace' is no longer fit for purpose if used in surface modelling that deals with decimals of a metre. If a contoured 'bulls-eye' in a modelled surface cannot be explained geologically, investigation may reveal that another Home Farm exists on the opposite side of the village, that there are two villages of similar name in the general area, each with a Home Farm, or that some other simple transcription error occurred when the data were first collected and registered.

These examples highlight the fundamental difference between verification and validation. Verification contributes a measure of assurance to the quality, reliability or usefulness of the pristine data; validation adds assurance that the quality of the data is not diminished during subsequent use.

## The significance of quality labelling

Good data management is seen more and more as a requirement in organizations that attempt to exercise quality control (QC) and aspire to offer services or products that are quality assured. It may not be appreciated that QC is a means to ensure that products or services meet required, agreed standards, not (necessarily) to ensure achievement of the highest possible standards. Quality assurance (QA) procedures include the considerations and activities that allow the definition and fulfilment of specified standards and assure the user that QC measures are being implemented effectively.

QC methods and their supporting QA procedures are themselves instances of fitness for purpose. For example, it might be agreed policy for an organization to respond to enquiries within three weeks and to operate working practices that achieve this. However, the quality of the response may be no higher than that from another organization that offers and achieves a response within three working days. The service offered by each organization is of high quality if the stated targets are achieved, and the user must decide which level of service is desirable. Generally a higher level of service, in this example the faster turn-around of enquiries, is more expensive and, on this basis alone, a slower but cheaper service might be preferable and perfectly acceptable to some users.

Quality management issues are intruding more and more into the activities of geological data managers, particularly those with tasks spanning the links between the materials, paper and digital formats and those dealing with data for use in digital modelling operations. An organization must address two major problems if planning to implement 'total' quality management (TQM).

- Working practices within different arms of the organization are adopted and adapted to optimize data collection, analysis and interpretation, to deliver specified products. To compare quality between different arms may be impractical, and potentially undesirable, if each arm uses methods and delivers products that are fit for specified purposes. The organization as a whole cannot claim to exercise TQM unless the practices, functions and performance of each quality system component are separately documented, monitored and regularly reviewed.
- More problematical are the linked questions of attempting to define and enforce quality management criteria retrospectively, in the face of unrecorded past working practices, or attempting to apply them to current working practices that are constantly evolving. In BGS, for example, it has to be admitted that achievement of the former is impossible and attempted achievement of the latter will be time consuming and expensive.

Fortunately, pragmatic alternatives are available. Robson (1994) pointed out that the aim of the BGS, a long-established and multi-disciplinary organization, should not be to conform to a single, all-embracing standard, but rather to apply a 'data qualification' (DQ) system to existing and newly acquired information (Fig. 4). A single corporate QA standard would necessarily be complex to establish, difficult to police and a burden to maintain, particularly with regard to 'historical' data holdings. It would include so many variables and caveats as to be meaningless. In contrast, DQ can be applied to selected archival information and newly collected data as and when required or when resources are available. DQ, as described by Robson (1994), is most obviously and most readily applicable to digital datasets, within which quality 'flags' would indicate levels of precision. Similar flags, covering individual data items or complete datasets can equally well be applied to material and paper data holdings.

There are, however, inherent difficulties in applying such quality levels to data in retrospect, particularly in situations where the lineage of the records is unknown or has become obscure due to lack of documentation. Another example can be drawn from the BGS borehole record system.

Borehole start point elevations, relative to ordnance datum (OD) in the UK, are vital to geological modelling. The relative reliability of each datum can be weighted according to the apparent 'accuracy' of its OD value. The BGS digital Borehole Database accepts OD values accurate to the nearest centimetre, if this fineness of measurement is recorded. The accuracy of the quoted value is then qualified against a set of four benchmarks:

1 (estimate only);
2 (accurate to nearest 1 m);
3 (accurate to nearest 0.1 m);
4 (accurate to nearest 0.01 m).

Working from modern logs that include reliably levelled OD values, this provides a useful key to the margins for error allowable during modelling, though the quoted value must be taken on trust. The system may break down when old data are transferred to the digital system by inexperienced operators or operators who are unaware of the history of the paper records. A contrived example based on a real problem illustrates this situation: operators reading an OD value of 21.336 m on a paper record might assume meticulous levelling had taken place at the time of drilling. An OD value of 21.34 m would be inserted in the digital system, together with an accuracy flag of 4 (accurate to nearest 0.01 m). If unaware of the data lineage, the operator could not appreciate that the seeming precision is illusory. The value is an artefact of a BGS-wide metrication exercise

carried out c. 25 years ago, when application of a conversion factor transposed a recorded value of 70 feet to 21.336 m. Reassessed in this context the accuracy flag must drop to 2 (accurate to nearest 1 m), since a single foot (0.3048 m) falls outside the limit of accuracy flag 3 (accurate to nearest 0.1 m).

The problem does not end here. The origin of the original 70 feet value may be unclear. Older Ordnance Survey maps showed contours at 25 (or sometimes 50) foot intervals and the recorded value could represent an interpolated estimate relative to the 50 and 75 foot contours or, with still less reliability, relative to 50 and 100 foot contours. An apparent accuracy of 4 is reduced to accuracy 1 (estimate only), and the fitness of the record for detailed modelling is seriously diminished.

If borehole data were transferred to the digital system by inexperienced staff without adequate instructions covering such problems, the value of the digital dataset would be undermined. This reinforces the need to employ well-trained, appropriately qualified staff, supervised by data managers who know their data and can anticipate the uses to which they will be put.

Implementation of DQ is therefore not without its pitfalls and requires time and effort to manage and accomplish. In organizations such as BGS, adoption of a DQ system would be much less costly and more meaningful than the establishment of wholesale QA accreditation. Also, DQ does not disallow local, project-based, quality systems within the corporate structure (Fig. 4).

## The 'intelligent geologist' concept

The importance of involving staff with appropriate background, skills and motivation in data management and related tasks has been raised above. Good data management practices and associated elements of quality management need realistic support in the form of adequate staff resources and realistic prospects of career development for those involved. Appropriate skills must be gained and maintained, underlain by a motivation to make the most of the data holding in its broadest sense. The data manager should exercise or impose rights of ownership over the information being managed, encouraging its improvement, but preventing inappropriate use, corruption or abuse.

A recent BGS project has delivered the Digital Map Production System (DMPS), in which geological maps are created from a digital database of geological attribute data. Early in

the project it became clear to those dealing with
the attribute data that an essential part of the
system was unquantifiable. This abstract element,
referred to by the DMPS team as 'the intelligent
geologist', represents a concept rather than a
single, gifted individual. Similar situations may
exist in other organizations, where skilled staff
apply accumulated experience and judgement to
make essential choices when faced with contradictory or equivocal data. Tacit acceptance that
the 'intelligent geologist' concept encapsulates
decision making and processes that are invisible
within the finished DMPS product, enables a
high-quality label to be placed upon each map,
even though the decisions and processes are not
quality labelled individually.

From a range of information, including first-
hand observations and measurements, earlier
work by other geologists and borehole records
of varied reliability, the 'intelligent geologist'
produces a rationalization that comprises the
approved geological map. This is a schematic
view, an *interpretation* of available facts using a
geologist's expertise. Data and interpretations
are adjusted to the resolution of the final map,
making best use of the relevant format, conventions and specifications.

The map produced by the geologist is a high-
quality product; but only one possible interpretation of the data available at the time of its
compilation. Within that product, individual
data items are not labelled for quality, though
some may have been labelled in the datasets from
which they derive. In preparing the map, the
geologist exercises a personal quality assessment
procedure; parts of this may be documented in
passing, but most of it happens instinctively and
subliminally. The map produced by capture and
reproduction of this analogue dataset in the
digital system is validated against the original,
but no attempt is made, nor can be made, at this
late stage to verify the data held within the digital
map database.

## The useful life of geological data

One of the greatest problems facing the
geological data manager is the assessment and
control of the useful life of data items. Data
managers in other fields find such tasks less
problematical. A common example, in a non-
geological context, is the way that a digital
system holding records of airline flight bookings
can be purged (or archived) when the flight is
safely completed.

Geological situations are rarely so clear cut.
Data apparently superseded by more recent

information or re-interpretations may still have
potential value. Modern mapping at BGS is
carried out upon the most up-to-date base maps
available. However, older map editions covering
the same area provide valuable information
about, for instance, the past extent of quarries
or former stream courses that have since been
culverted. Similar examples are common and
cover the full width of geological data holdings,
from draft texts to tables of interim analytical
results. For instance, raw data used to calculate
a radiometric date cannot be assumed to be
valueless once the date is published. Later
development may indicate that inappropriate
constants or equations were used. Availability of
original data allows re-calculation without
costly re-analysis.

A limited number of discrete, perhaps site-
specific datasets might exist within the geological
sciences that have a short life and no apparent
re-use value. Generally, however, the potential
lifetime of geological data, whether archived or
active, is unlimited. Data collections must be
stored, monitored and their usefulness reviewed,
and it is the question of 'when to stop' that
remains most difficult to answer.

## The growing need for data management

The above discussion is an overview of some
major issues and problems faced by the modern
geological data manager and describes some of
the expedients that may be adopted to solve
them or minimize their effects. There are
compelling scientific reasons for operating
good data management practice, but until
relatively recently, the will, the support and the
means have generally been unavailable in many
organizations (*cf.* Feineman 1992).

Modern digital systems are able to store and
manipulate information in quantities and at
speeds that were inconceivable 25 years ago.
Some digital operations generate new data and
these new data may readily be edited to provide
yet more data. Management of data in digital
systems has many facets. Some management
occurs wholly within the digital system and it
must be decided whether interim or superseded
versions are archived, and if so whether they are
archived digitally on one of a variety of media,
or whether they are 'dumped' back onto paper.
It may be necessary to archive 'history' data and
these may also spawn paper data. Sometimes a
lengthy, multistage, digital operation, such as
the digital capture and attribution of geological
map information, will lead to production of
offspring digital data such as plot files, or

derived paper data such as the eventual map plot. Labelling, tracking, protecting and backing-up this multiplicity of information diligently and efficiently is time-consuming and, hence, expensive. When viewed in terms of the costs of replacing data, the potential value of data holdings and the income that good-quality data can generate, such expense is justified.

As the use of digital data has grown in importance, technology has developed to support their storage and management. Attempting to transfer unmanaged or poorly managed analogue information to digital systems of increasing functionality and sophistication is fraught with difficulty. Attempting to take existing analogue information at face value and converting it to digital form for storage or manipulation in computer systems highlights three things.

- Once the information enters the digital realm it acquires a level of apparent authenticity that is commonly far beyond that which would be attached to the original data. This situation will improve as users accept that mere transfer of data to a digital system does not improve data quality; it only increases the ability to store and manipulate the data.
- Digital systems are less forgiving of inaccurate, incomplete or illogical data than are equivalent manual systems (which in the BGS context include the 'intelligent geologist'). This second point is one of two keystone factors that have raised awareness of the need for geological data management.
- The second keystone is more readily apparent from the considerations above. If management of data in digital systems requires explicit rather than implicit quality management, it must be accepted that, logically, similar quality management and supporting data management are just as important to non-digital information.

This last point might appear less urgent to those who have worked through an organization where poorly documented manual systems have appeared to succeed due to the power and adaptability of the human brain (e.g. the BGS 'intelligent geologist'). On this basis the need to address data management problems in non-digital areas might have continued to receive only notional support, or no support at all. However, the factor of QA or DQ has now entered the equation.

In the modern, competitive world, most organizations can no longer afford simply to get by. Nor can an organization that wishes to prosper and remain viable be prepared to live on its reputation or survive on reserves of data and expertise put by during earlier, easier times. The adage that you must 'speculate to accumulate' has taken on new meaning in the context of data collection and its onward use.

It has become vital to recognize the quality and know the location of data holdings, and to offer products that trade upon the perceived quality of the raw and derived data and the expertise involved in collecting, producing and interpreting them. In the BGS context, products include traditional geological maps and descriptive memoirs, but more and more customer-specific output, such as thematic maps, geological models, or expert advisory services on a multitude of topics, is required.

To compete, organizations offering 'quality assured' products must address the gamut of quality management issues and ensure that their own products and services are 'fit for purpose'. The BGS is one of many organizations working in the geological field, all of which must upgrade their data and quality management activities if they are to remain competitive. The corollary of this is that in carrying out this upgrade to ensure competitiveness, the organization will also initiate an overall increase in the quality of their products and services and ensure that new, potentially valuable, data continue to be collected and managed.

## Conclusions

There are areas within geological science where the need for data management has been underrated, due to preoccupation with data capture and data interpretation for scientific research or commercial investigation. The need to look after data, to understand their limitations and usefulness and to ensure their availability seemed of secondary importance. Data managers (commonly regarded as clerks) were undervalued, their skills unrecognized and their prospects, limited. In enlightened organizations, such is no longer the case.

This paper opens with a quote from a report (Woodland et al. 1973) that drew attention to the problem of managing geological data. The intervening years saw little improvement of the situation. Only recently has a combination of factors revolving around the issues of data quality and data value forced a reconsideration of priorities and data management policies. Among the recommendations and conclusions addressed to the executives of the Natural Environment Research Council ('Council'), Woodland *et al.* (1973, p. 7) re-emphasized that: 'The environmental sciences amass large quantities of diffuse data, the successful processing of which will require adequate systems of data management. This has so far been a largely neglected field where predominantly qualitative data are concerned.' In addition, the authors requested: 'Council is asked to encourage developments leading to the provision of adequate data management programs for all branches of geology.'

These observations may have been noted and possibly received support in some areas of geological activity. Elsewhere reaction has been slow, limited or non-existent. Awareness of the need for geological data management outlined in this paper is, in many respects, due to the demands of mature digital technology, which was in its infancy when studied by Woodland's working party.

It is gradually being accepted that a real need for efficient data management and quality management exists in the geological sciences. These needs will not be satisfied cheaply or casually; the role of the geological data manager must necessarily expand. As the role expands, increased recognition of the data managers' experience and skills, and the benefits that their efforts can accrue, should lead to a further raising of the profile of, and support for, geological data management.

My thanks to colleagues in the Data and Digital Systems Group of the BGS for their helpful suggestions and to the many staff who have worked on the Digital Map Production Implementation Project. Their expertise has opened my eyes to many previously unperceived data management problems. Roy Lowry and an anonymous referee read the draft paper and their constructive comments have undoubtedly led to improvement of this version. Views in the paper remain my own, the perceptions of a geologist 'poacher' turned data manager 'gamekeeper'. Others will see the problems, their solutions and the role of the geological data manager, differently. The work is published with the approval of the Director, British Geological Survey (NERC).

## References

FEINEMAN, D. R. 1992. *Data Management: Yesterday, Today and Tomorrow.* Presentation to the PETEX '92 Conference, 25 November 1992, London.

ROBSON, P. G. 1994. *Quality Assurance of Data in the BGS.* British Geological Survey Technical Report, **WO/94/11R**.

WOODLAND, A. W., ALLEN, P., BICKMORE, D. P. ET AL. 1973. Report of the working party on data processing in geology and geophysics. *In: Research in the Geological Sciences: A Consultative Document Comprising Reports from Eleven NERC Geological Sciences Working Parties.* The Natural Environment Research Council Publications Series 'B', No. 7, 1–7.

# The quality assurance of geological data

## PAUL R. DULLER

*Amerada Hess Ltd, 33 Grosvenor Place, London SW1X 7HY, UK*

**Abstract:** The key to success in any geological investigation is access to accurate and reliable information. Clear strategies are required to ensure that geological information, often acquired at great expense, is not mislaid and is kept in a form that can benefit its owners. To meet this need requires the provision of a reliable data management strategy, a well-defined and efficient policy of administration and a clear understanding of the nature, origin and quality of data involved. Geological data by their very nature pose particular problems in terms of compilation, quantification and validation. However, the application of standard quality control, quality assurance and audit procedures can be used to ensure the *correctness* (or accuracy) of the data, thereby safeguarding the integrity of any associated investigation.

Geoscientists are in a relatively unique scientific position given the vast array of data sources and analytical techniques that they have access to. In addition, developments in information technology and remote sensing over the last decade have revolutionized the manner and speed with which they can now collect, integrate and manipulate geological data. It must be recognized, however, that the success of any geological investigation depends upon access to accurate and reliable information. This paper provides an overview of the processes involved and actions required to achieve this.

## So what is the problem?

In any geological study it is now quite normal for a large volume of data to be assembled from a diverse range of sources and integrated to form the basis of a geological model. The rapid increase in the use of computer systems for the management of geological data over the last decade, has greatly helped the data collection and validation process associated with such studies, and databases containing millions of characters are now commonplace. The problem now facing the geological community was aptly summarized by Openshaw (1989): 'The ease with which data from different scales ... with different levels of innate accuracy can be mixed, integrated and manipulated, totally disguises the likely reality of the situation.' In particular, as few geologists understand or have quantified the nature of errors associated with their data, the quality of geological information is often taken for granted. It is important to realize that unless the geological datasets have been compiled under properly supervised and controlled conditions, verified, validated and audited, then their quality cannot be assured. Poor-quality data by their very nature can lead to incorrect, if

not misleading, conclusions and costly mistakes. It is for this reason that users of geological datasets and database management systems must have confidence in both their reliability and data quality. All geological datasets should therefore be compiled by professional staff, quality assured and audited before use.

## Sources of error

The term 'quality' is the subject of widespread use and abuse. Although quality can be a hard concept to define, when it is absent, it is highly conspicuous (Price 1984). Errors form the most frequently cited elements of data quality and can arise at any stage during the life cycle of a data item. Errors may derive from many sources and may relate to data collection/compilation processes, variable precision or accuracy, the timeliness of the data, reliability, consistency, completeness, transposition, and/or data input errors.

## Other factors

A number of other factors may have a direct effect upon data quality yet are not immediately obvious. These include the area or coverage of the dataset, its positional accuracy, original purpose, analytical history, errors in the original source material, the existence of two or more possible data sources, data processing errors, loss or lack of original data, and administration controls such as security and access.

## Who needs a data management strategy

Clear strategies are required if any organization or individual wishes to ensure that data, often

*From* Giles, J. R. A. (ed.) 1995, *Geological Data Management,*
Geological Society Special Publication No 97, pp. 91–95.

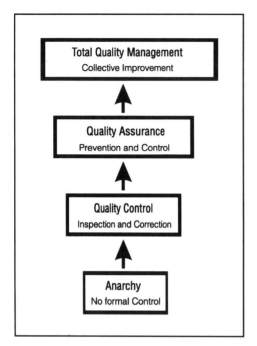

**Fig. 1.** The road to quality.

with their compilation rules, quality control procedures and potential sources of error. Data *users* are expected to understand the content of the data, contribute data and add value to the data. *Administrators* do not necessarily have a detailed understanding of the significance of the data, but have a detailed record of (and the authority to allow) access or additions to the data. The administrator analyses relationships in the data, controls the addition of new data attributes, prevents unauthorized additions or changes, provides a reporting function and assists users in identifying what data exist and where they can be found. The *curator* is responsible for the physical well-being of the data, checking that the file of data is on the right shelf, disk or tape and is secure from unauthorized use or theft. In addition to these roles, a much deeper collective role for the organization as a whole is of fundamental importance. The *collective user* role is based on the premise that if the data management system is not trusted it will not be used, and that if it is not used it will become unusable. It is therefore important to realize that a properly managed database in which the various items of data are consistent and accessible, is of far greater value than the basic data catalogue itself.

acquired at great expense, are not mislaid and are kept in a form in which they can be used to the benefit of their owners. The ability to access *quality assured* data held throughout an organization and present them in a compatible, comparative and meaningful manner can substantially improve the quality of any geological investigation and may identify trends and patterns not previously apparent. Many simple procedures can be set up to record or improve data quality and form the basis of such a data management strategy and quality system. These include the establishment of data dictionaries; compilation of metadata; development and documentation of validation rules; design of quality assurance procedures; use of professional staff for verification/audit; quality assessment and labelling of existing datasets and the assignment of data ownership responsibilities for all data attributes and records to specific users (see Fig. 1).

## Key roles in data management

It is convenient to distinguish four key roles relating to data management, namely: collectors, users, administrators and curators. The *collectors* require a detailed understanding of the source, nature and origin of the data, together

## How do we add quality?

It is one thing to collect data and quite another to know that all the necessary data required for processing have been correctly compiled, input, validated and verified. The processes of validation, verification and audit may be grouped into three distinct tasks: the identification and definition of potential sources of error; the design and implementation of the necessary quality control and quality assurance procedures; and regular audit/inspection of the data by professional staff.

### Validation and verification controls

The objective of data validation is to detect errors at the earliest possible stage, before costly activities are performed on invalid data. It is therefore essential to ensure that source data are correctly recorded, before data preparation takes place. Similarly, it is important to check the accuracy of data preparation operations before data are processed, which is achieved by verification procedures. Unfortunately, the historical nature of some geological data poses particular problems to validation and error correction. For example, although specific attributes such as chronostratigraphy may readily be defined, coded and dictionaried, lithostratigraphical usage differs between organizations, regions, basins and within its own historical context. Similarly, many other forms of geological data may be difficult to validate automatically and must be subject to stringent visual auditing and cross-reference procedures.

When data are input they can be subjected to a vetting procedure by means of interactive validation routines that may be used to *interrogate* or check individual data entry fields and *trigger* a specific response when an attempt is made to enter invalid data. Whenever possible, data-entry systems should be designed so that they are self-checking and self-correcting. It is therefore important that data compilers, users and administrators are consulted by system designers to establish the types of checks and controls required to ensure that incorrect or invalid data are rejected. During the various input stages several types of check may be performed to ensure that the data are of the correct type, relate to the correct sample/location, are set out in the expected format and conform to the minimum and maximum range of values. During the initial data compilation process, the compilation teams should validate codes (by reference to a data dictionary), ensure the compatibility of two or more data sources and avoid unnecessary rejection of data due to purely random causes.

## Quality control versus quality assurance

The quality of any dataset can be increased by the application of simple quality control, quality assurance and auditing procedures. Price (1984) defined quality control as consisting of 'little more than measuring things, counting things and doing something simple and sensible with the results of the measuring and counting....It is easily learned, there is no mystery in it, and certainly nothing clever.' Quality assurance can be viewed as an attempt to separate inherent and assignable variability, by posing four questions about any given product, dataset or attribute:

- Can we make it OK?
  (also known as process capability analysis)
- Are we making it OK?
  (the quality control/validation mechanism)
- Have we made it OK?
  (the verification/audit processes)
- Could we have made it better?
  (research, development and process evolution)

There are three basic rules of quality control that may be applied in any environment, geological or otherwise, namely: no inspection, measurement or audit without proper recording; no recording without analysis; and no analysis without action (even if the action is to recommend no further action).

Quality Assurance (QA) represents a formal demonstration of the systematic and committed management control of all things that affect the quality of our data. The difference between quality control and data auditing procedures is that an audit is usually undertaken after a significant part of the data compilation and input processes have been completed. If any systematic errors are detected, it is usually too late to put them right and as such, the data gathering processes must be revisited. Quality control methodology, however, involves the inspection of a series of pre-defined samples during compilation and input. If things are going wrong, this is detected, corrected and prevented.

## Metadata

A powerful tool in the analysis and implementation of quality standards is the use of *metadata* (Ralphs 1993). This is the term applied to data held in additional fields, containing comprehensive documentation on the datasets themselves. Metadata may include: information on the source, nature and origin of data, its assessed quality, processing history and precision/accuracy, and notes on standards, rules and procedures used in the data generation and validation and verification processes.

## Documentation and procedures

Documentation is vital. It can take many forms and may be as simple or complex as is deemed necessary by the data administrator. Procedures form the backbone of any formal quality assurance system. Their purpose is to communicate information and provide clear guidance to people performing tasks. The need for good communication is always important, particularly when activities are both complex and diverse. Unless processes are properly defined and documented, people will of necessity do things their own way to their own standards. When the latter happens the outcome, despite their best intentions, will invariably be limited overall effectiveness, duplication of effort, varying standards, wasted time and resources, and poor-quality data.

The drafting of a procedure is very important and should be given careful consideration. Never forget that what is written will define the manner in which and the standard to which people are expected to perform. It provides a unique opportunity to get things right, to become both effective (*doing the right things*) and efficient (*doing things right*). However, it is not necessary to tell trained and experienced people how to do their jobs. Instructions should not set unrealistic standards nor reflect purely personal foibles.

Documentation should include details of data sources, compilation rules, codification details, validation rules, consistency checks, data ownership, workflow (i.e. the periodicity of inputs/outputs, data transfers, etc.) and both user and programmer documentation for each in-house data input/processing system. All documentation should contain the date and version number on each page, in order to quality control the document itself. In addition, a catalogue of deficiencies and problem areas within existing systems and datasets should be compiled in order to allow prioritization of further development and maintenance work.

## Quality management systems

The development of a formal quality management system starts by identifying in a policy statement the things that affect data quality. This policy statement is then supported with a quality manual and procedures stating how the policies are achieved. Procedures may be complimented by written instructions and are supported with internal audits and spot checks to provide proof of implementation and compliance. There are several ways of generating a quality management system, ranging from trial and error through to employing a quality management consultant. Whatever the method selected, a number of features are important:

- commitment and leadership from senior management;
- management awareness and understanding of the subject;
- quality management training and participation at all levels;
- clear purpose and objectives;
- effective response to shortcomings;
- a programme and budget.

**Fig. 2.** Lack of forward planning!

*Quality standards*

Many countries have professional institutes that deal with quality control policies on both a national and an international scale. In the UK the British Standards Institute has established a recognized quality management standard (BS 5750), which forms the UK equivalent of the widely accepted European (EN 29000) and international (ISO 9000) quality standards. To reach this quality standard, any organization must follow several predefined stages, including the documentation of all procedures used by each unit or division; an internal acceptance of these standards; and periodic audits by an external peer group to ensure that the defined internal standards meet the required needs of the organization and are properly used.

## Conclusions and recommendations

It is clear that geological data are being collected and distributed faster now than ever before. The challenge now facing the geological community is to admit that data errors exist in practically all datasets, and to consider the limitations that the errors impose rather than ignoring them. Although technology will never replace the human quality control process, an awareness of the problems and possible solutions associated with geological datasets should be of benefit to everyone. To avoid propagating data quality problems the following procedures should be followed:

- document and review any existing data management strategy;
- develop new methods, procedures and quality standards;
- assess the suitability of disparate datasets before their integration;
- stress and propagate the strategic importance of data quality at all levels within your organization;
- identify *data owners* for each attribute and data record within your database;
- develop the role and improve the status of the data administrator.

New technology has often been brought in to troubleshoot situations where quality controls have broken down or never existed, but used in this way, it is only a temporary remedy for immediate problems (see Fig. 2). The need for a complete change of attitude towards forward planning and management, to anticipate the demands and limitations of geological data before it is too late to do anything about them is increasingly apparent. Within any organization, quality assurance will only work if understanding, commitment and support are visible at all levels of management and transmitted through proper communication, training and procedures to everyone involved in any aspect of data management.

## References

OPENSHAW, S. 1989. Learning to live with errors in spatial databases. *In*: GOODCHILD, M. & GOPAL, S. (eds) *Accuracy of Spatial Databases*. Taylor & Francis, London, 263.

PRICE, F. 1984. *Right First Time: Using Quality Control for Profit*. Wildwood House Ltd.

RALPHS, M. 1993. Data quality issues and GIS—A discussion. *Mapping Awareness and GIS in Europe*, **7**, 7, 39–41.

# Project databases: standards and security

## STEPHEN HENLEY

*CSIRO Exploration & Mining, 39 Fairway, Nedlands, WA 6009, Australia*

**Abstract:** For technical applications in the mining industry, large institutional data models are the exception rather than the norm. Much more common, given the distributed nature of the business, and the scattered locations of the specialists involved, is the personal or project database. Such a database is set up for a project (or part of a project), and is developed using a data model defined specifically for that project. At the end of the project, the database may be archived or it may be supplied to a central group for integration with a corporate database. If uncontrolled, such methods of database management can lead either to direct drastic loss of data or to incompatibilities in data recording standards. The use of common software standards and data dictionaries throughout an organization can reduce radically these dangers and can produce positive benefits not only in data recovery but in reusability of techniques.

## Introduction

Much is said about the rapid increase in the rate of acquisition of data. However, much less is heard about the increasing rate of loss of data, which goes almost completely unnoticed except by those who are directly affected.

There are data lost by physical degradation of paper or magnetic media: such losses can be reduced, though not eliminated, by regular copying or back-up on to new media. There is also large-scale loss of data often from entire projects because it cannot be located, or when located it cannot be read or understood.

Many attempts have been made to prevent such data loss by the integration of all data into monolithic corporate databases. These have usually failed or at least have been only partially implemented, because of a lack of individual commitment, or a loss of corporate interest during the inevitable delays in implementation, or lack of the necessary management structures and incentives. However, the principal reason for such failures is probably that few large organizations can demonstrate a tangible short-term benefit to be obtained from the creation and continued use of the necessarily complicated databases.

In recent years, data held on PCs have been particularly vulnerable to loss because of the fragility of back-up media and because of the frequent absence of corporate priority and policies for the preservation of project data. This is so, particularly in the commercial world: data from an exploration or mining project has no value once the project is either abandoned or completed, and it is often difficult to justify expenditure on recovering project data into any central repository. Typically such commercial project data have a life expectancy of only a couple of years after the end of the project.

## The corporate database

Over the years, many larger organizations have tried to develop unified technical databases to cover their entire range of data, across divisions and departments. This was particularly so in the days of predominant central mainframe computer services, but still continues today. The investment is large, and the database is necessarily seen as long-term.

One of the essential features is the development of grand 'data models', such as that proposed in 1993 by Ken Rasmussen of Logica, for use by the British Geological Survey, and that proposed by POSC (the Petrotechnical Open Software Corporation) for use by companies in the oil industry.

All users in the organization are expected to follow the database standards, and even now are expected to be connected on-line to a shared central facility; there is a necessity for strict and complex database administration procedures to avoid chaos. The software used for access to the database is also shared, though applications software may be diverse and held locally for operation on retrieved datasets.

Many such pre-defined data models have failed because they lack the incentives to users to adhere to their strict requirements, and because they lack the flexibility needed by users to handle idiosyncratic datasets in the most efficient way. The likely benefits from use of such data models are not apparent to everyone at the outset, and it is unclear to most staff responsible for building the databases what benefits they, individually, will ever see. What they can see is additional workload which is tangential to their own job description. Nevertheless, they provide a secure environment in which data integrity and longevity is largely ensured by the imposed central disciplines and regular back-up procedures.

*From* Giles, J. R. A. (ed.) 1995, *Geological Data Management,*
Geological Society Special Publication No 97, pp. 97–101.

## The project database

The project database is perhaps the area of most significance, because this is where most databases start from, even if at some point the data are transferred into a corporate database.

In organizations large and small—geological surveys, universities, and commercial companies of all sizes—data are collected and processed on the project scale, by individuals or relatively small teams. As soon as a team grows to perhaps 10, 20 or more individuals, there is a tendency to partition responsibility so that databases become subdivided.

There is certainly commonality among projects in very many cases. Often there are standard data dictionaries in use organization-wide, in the same way that some standards are defined nationally or internationally. However, the task of maintaining the project database rests with the project manager or whoever he or she might delegate. The project database, whether held on a local computer (PC or graphics workstation) or in a shared host computer, usually remains for the duration of the project under the control of the project team. This is a matter of necessity and common sense, as until the project is completed there is no certainty that data will have been completely validated, let alone be complete.

Project lifetimes tend to be much shorter than organizational lifetimes, though quite often (such as in mining consultancy) a project team may be put together outside any organizational framework.

Data used in a project may reside in a very wide variety of software environments, and there are not necessarily any connections among some of them. For example, there may be no good reason why financial data for a project should be integrated with geological exploration data or process design data. Three completely different systems are used for processing these data types, and the connections among them can be relatively tenuous.

There will also be a strong likelihood that data are held on separate computers. This is a natural consequence of project-based management; without strong central management there is always the tendency for a project to try to be self-contained. Even if connected to a network for what data it can receive, this will tend to be a one-way traffic.

The software used may be diverse. Again, this is a consequence of project management. The project staff will wish to use software, including databases and spreadsheets, that they have experience in and are comfortable with. It will differ

with the project, and will not necessarily have the seal of corporate approval.

At the end of the project, then, there will be a body of data held in a heterogeneous set of files in miscellaneous formats. This will include raw data still much in the form that it was collected, perhaps with some editing to remove coding errors. It will also include files of processed data, constructed by combination and analysis of the raw data files, and varying in significance from scratch files to the important end results of the project. A special case of processed data is files containing data for plotting; graphics metafiles, for example.

There will also be a collection of macros or batch command files which have been built up during the project, to obtain useful interpretations, plots, and reports from the various types of data. These files in themselves are an important resource. There may also be, usually on paper rather than in machine readable form, some assessment of the quality and significance of the different datasets. This could include such items as notes on analytical techniques used, or comments on currency stability or commodity price forecasts for financial projections.

What happens at the end of the project? In a well-run organization, of course, such data are always transferred to a corporate database, after due scrutiny by the database administrators.

As a second best, the project database may be off-loaded to diskettes or to back-up tapes, hopefully in more than one copy, but perhaps in compressed formats. These may, with luck, be kept centrally in a clean air-conditioned computer room or storage facility (perhaps a second best to being put into the corporate database). Of course, if the project used network access to central computing services, there may be no need for the project manager to do anything—it may just be assumed (or hoped) that the computing service takes care of everything.

In most organizations, the likelihood is that such project data actually finish their days in box files; and in any case the significance of interpretations diminishes with time and the disappearance of staff who actually remember the project. And then, what happens to the box files? With luck they will gather dust in a central library or archive. More commonly, within a surprisingly short time after the end of a project (years or even months), other destinations may be found.

After the end of a project, the subsequent lifespan of data can then be very short, and data can be lost in a wide variety of ways.

## Physical loss

Physical loss is the most obvious and dramatic. Fire and theft are clear causes of loss, and care is often taken to forestall these through the obvious security measures in the case of theft, and the normal fire precautions. However, it is often forgotten how sensitive computer media are to both moisture and dust, as well as high temperatures. A fire which is extinguished long before it guts the building can still destroy the data on a lot of tapes and discs. But, less dramatically and much more commonly, data may also be misfiled, mislaid, or just plain lost. Of course, there should never be only one copy of any set of data, but even if two copies are kept, the loss of one may not be compensated by the production of another copy; even if it is, that is the time when it may well be discovered that the second copy has become corrupted.

## Degeneration of media

Discs and tapes are subject to the normal processes of degradation subject to temperature and humidity changes, and atmospheric dust. Unless stored in a clean air-conditioned environment, the life of a disc can be surprisingly short, especially if care is not taken to keep it free of dust and protect it also from stray magnetic fields.

In the 1960s and 1970s it became standard *commercial computing* practice to keep multiple copies of magnetic tapes. However, in the past ten years, with rapid increase in capacity and reduction in price of hard disc drives, much of this discipline has been lost. It is common (though unwise) to rely on a single disc drive for the original and back-up copies during database updating. Yet the risks of trashing a hard disc, either by failure of an electronic or mechanical component, or as a result of a software fault, are such that it is unwise in the extreme not to make frequent back-ups. Diskette capacity has not kept pace with the hard drive, thus making this a problematic back-up medium, but it will be some time before the hard disk totally outstrips DAT (digital audio tape), currently perhaps the most convenient and cost-effective back-up medium.

However, all magnetic media, including the original standard nine-track tape, are subject to gradual deterioration, and it is a wise policy to make further copies at periodic intervals, giving the opportunity also to transfer data to more up-to-date media.

## Change in technology

The changes in technology are most insidious. There have been five major changes in storage technology in the last 25 years. With time it has become progressively more difficult to access data on paper tape, punched cards (there is still to this day a card store at Mufulira, Zambia, with many tens of thousands of punched cards containing important geological data), on nine-track tape, and on 8-inch floppy discs. There are many old $5\frac{1}{4}$-inch floppy discs which are unreadable today because they were produced on CP/M microcomputers in a huge number of different formats. Today, the $5\frac{1}{4}$-inch floppy disc is still in use but has been overtaken by the 3.5-inch diskette. Although both may seem entrenched for all time to come, the coming of the CD drive suggests the day may not be far off when all magnetic storage media will be consigned to history. Data stored on such media may simply fade away through increasing difficulty in finding any way to read them.

## Change in software

If data are encoded in ASCII (or even in BCD or EBCDIC) there is little if any problem of software dependence, though the format and any codes used must be fully documented. However, most software systems define their own standard formats, and these change with time. Furthermore, if using a data compression utility, the formats used by these also change with time. For example, for those who use PKZIP, any files compressed using the 2.04 release of 1992 are unlikely to be readable by earlier versions. This is not usually a problem of course, until the program itself falls into disuse. The program and version used for compression should be identified, and ideally stored with the data itself (for example, in this case by using the ZIP2EXE program which converts the compressed file into an executable which will unpack itself).

For those in the British Geological Survey and elsewhere who have used G-EXEC (their standard geological data handling system for most of the 1970s; Jeffery & Gill 1976, 1977) in the distant past, it is salutary to consider, however, how it might be possible now to recover data held in a G-EXEC standard binary file. The same is true of many other packages: many consider now that any of AutoCAD, or dBase, or Lotus 1-2-3 provides a convenient format for data storage and interchange. However, even these formats are subject to change and will sooner or later be abandoned. Looking more specifically at geological software, we fare little better. Database formats have changed in all the leading products. For example, DATAMINE (Henley 1993) originally had a database structure in which logical data files were mapped into fixed-format fixed-size physical files. Logical data files are now stored separately. It is an open question for how much longer the old file format will be supported, and how those databases (if any still exist) will be accessed in the future.

### Loss of paper records or indexes

The loss of paper records or indexes describing the data formats and contents can itself lead to loss of the data. Even if the media can be read, and the files can be recognized, loss of paper records describing the project itself and the significance of the data can render the data totally useless.

### Loss of corporate memory

Loss of corporate memory is the inability to locate the data, even if the media may still exist and be readable.

Some of these problems are easily solvable, with the right management approach. Others are intractable with any approach but their effects can be ameliorated by suitable procedures.

### Database standards

The world of general database management systems is evolving very fast. Even if one adopts today's market leader, there is no guaranteed lifetime, and a very strong likelihood that in five years' time it will no longer hold that position and that furthermore the product with the same name in five years' time will be completely different. Not all suppliers are committed to maintenance of full upwards com-patibility, and the necessary conversion utilities may be more difficult to obtain retrospectively than the systems themselves.

There have been many claims over recent years for future-proofed hardware and software systems, but none has yet stood the test of time. As a database interface standard, SQL has displayed commendable longevity, but it is not quite as standard as database suppliers would like users to believe, and in any case it unfortunately says nothing about storage standards.

The European Community's DEEP project, in which partners from five European countries are participating, is an attempt to define a standard set of methods for database access, for exploration and mining data. In this sense it goes well beyond the scope of SQL, and will provide the basis for much easier data interchange as well as for interoperability of software from diverse vendor organizations. This project should be completed by late 1995.

### Database integrity

The DEEP project has much to say about database integrity. It is concerned with data models in a generic sense. It is intended to provide a prototype of a system suitable for the corporate level database, but also for the project database. It will define a series of geological and mining data models for demonstration, but more importantly will provide a framework for definition of new data models. DEEP does not set out to define rigidly a complete range of computing standards like POSC, but rather to make available a new technology to geologists and mining engineers. This includes a framework for inclusion of the descriptive 'paper records' as an integral part of the database.

There is still, of course, a need for discipline within the organization, but with coherence between the corporate database (if any) and the project database, or among different project databases, life becomes much easier.

### Reusability of data

By adopting such strategies, a major benefit is obtained by the organization even if all data when collected are held only in project-specific databases. It becomes possible for data from different projects to be compared and combined. The useful life of the data is extended, as recording and storage of the data follow known standards. Even if the standards themselves may be changed from time to time, the problems, for example, of idiosyncratic and *ad hoc* coding schemes, are avoided.

The problems of physical loss and technology change *can* be minimized by adopting fairly simple data management procedures. However, this is always subject to cost–benefit analysis. The question that must always be asked is 'Is my database worth more than the time and expense of keeping it accessible?'

The answer to this question is normally 'yes' considering the cost of data collection and interpretation. But it is not an answer that can be imposed generally; each data user must answer it for his or her own particular situation. This paper discusses only the generalities.

## Conclusions

There have been many diverse approaches to database management. Those geological organizations which have, in the past, tried to set up and enforce corporate-wide database structures have found them expensive and unwieldy. Those which have allowed their staff freedom to build private databases with little or no central control have lost their data. A compromise approach, in which proper procedures for coding and for security are enforced, but which allow the creation and management of separate project databases, can allow and encourage the preservation and sharing of data. Data model standards, such as those being developed in the European DEEP project, are an essential component of such an approach.

## References

HENLEY, S. 1993. Computer applications in mining geology. *Geoscientist*, **3**, 20–21.

JEFFERY, K. G. & GILL, E. M. 1976. The design philosophy of the G-EXEC system. *Computers and Geosciences*, **2**, 345–346.

—— & —— 1977. The use of G-EXEC for resource analysis. *Mathematical Geology*, **9**, 265–272.

# Database applications supporting Community Research Projects in NERC marine sciences

ROY K. LOWRY & RAYMOND N. CRAMER

*British Oceanographic Data Centre, Bidston Observatory, Birkenhead, Wirral, Merseyside L43 7RA, UK*

**Abstract:** Traditionally, oceanographic data management has been strongly cruise oriented. However, the organization of research into Community Research Projects, such as the North Sea Project and Biogeochemical Ocean Flux Study (BOFS), has produced a requirement for handling multidisciplinary data from a large number of research cruises as an integrated dataset. The code of practice for the project dictated that data were to be exchanged freely between project participants throughout the duration of the project, after which the data were to be brought into the public academic domain. The data management strategy therefore had to provide both a vehicle for data exchange within the project and a mechanism for subsequently making the data available to a wider community.

These problems were solved by BODC through the adoption of two technologies. Relational database technology was used to integrate the data and, through access over the UK academic wide area network (JANET), provide the data exchange mechanism. Electronic publication of the datasets on CD-ROM provided the mechanism for expanding the dataset user base.

For many years oceanographers have put to sea to collect the data on which their science is based. Until the mid-1980s, the following pattern of working was used by most research groups. First, the research objectives were determined and a block of ship time, termed a cruise, was allocated. Each cruise was manned by personnel from the research group who would collect their cruise dataset, work it up back in the laboratory and interpret it in the light of the cruise objectives.

Data management philosophy in this scenario was firmly focused on data archival. In theory, once the dataset had been interpreted and the scientific papers written, the dataset was passed to a data centre for long-term archiving. All too often what happened in practice was that the data were put on one side to be dealt with later, and subsequently forgotten.

Scientific data which lie outside professional data management are extremely vulnerable. Often they are held without adequate back-ups on perishable media or on media which have become technologically redundant. Even data held on the disc store of a professionally backed up system can become vulnerable as computers are replaced. Consequently, many cruise datasets have been lost after the primary objectives for which they were collected had been fulfilled.

## Community research projects

In the mid-1980s, the Marine and Atmospheric Sciences Directorate of the Natural Environment Research Council (NERC) introduced the concept of Community Research Projects. In these, a team of scientists from several NERC institutes and university departments were brought together to work on a common problem. The projects were organized into a Core Programme which provided a supporting platform for a large number of grant-maintained research projects termed 'Special Topics'.

The result was large-scale science bringing together teams of over a hundred scientists scattered throughout the United Kingdom. Two of the Community Research Projects, the North Sea Project and BOFS, included a significant data-gathering component with approximately a year and a half of ship time allocated to each. BODC was charged with the provision of data management support to these two projects.

## Data management requirements

A new approach to the management of data was required for two reasons. First, the scale of the science raised the stakes. When the amount of resources directed towards these projects was considered, the dataset produced could not be allowed to evaporate into the ether. A strategy therefore had to be developed to ensure that the proportion of the data reaching the safety of professional data management improved dramatically. Further, to justify their costs such projects must produce 'deliverables'. Usually in science these take the form of published papers,

*From* Giles, J. R. A. (ed.) 1995, *Geological Data Management*,
Geological Society Special Publication No 97, pp. 103–107.

but a most welcome addition to these is a clearly identified project dataset.

Secondly, cruises within these projects differed markedly from those that had gone before and the status of individual cruises as discrete entities had waned considerably. For example, the North Sea Project field work consisted of 38 discrete cruises. These were manned by numerous permutations of personnel from the Core Project and several Special Topics with up to 10 different research organizations represented on a ship with 14 scientific berths. Consequently, there was a requirement for the data from all cruises to be brought together into an integrated dataset which had to be made available to the project scientists at the start of their interpretative work, not when they had finished.

A further problem arose from the distributed nature of the cruise personnel. When a single research group was involved in a cruise, working up the automatically logged data from the cruise presented no problems: the data were simply transferred from the ship's computer to the system used by the research group. However, consider the case of one of the instrument packages, the CTD (conductivity, temperature and depth), used in the North Sea Project.

Responsibility for the different channels was split between a laboratory in Birkenhead, Southampton University and two organizations in Plymouth. The logistics of data distribution and recombination of the worked up product were too horrific to contemplate. The data management requirement therefore had to incorporate a centralized data-processing facility.

These requirements were met by the data management model shown in Fig. 1. BODC personnel worked in close co-operation with the project scientists to process the data which were then loaded into the project database and hence made available to the scientists. The good working relationships established in this way greatly encouraged the flow of the scientist's own datasets into BODC. This proved a most effective solution in both the North Sea Project and BOFS.

## Project database

### Design considerations

The objectives of a project database supporting a Community Research Project require some

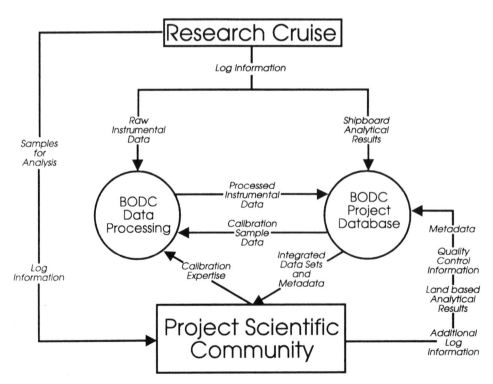

**Fig. 1.** Data management model adopted for the North Sea Project and BOFS Community Research Projects.

examination. There are three important aspects. First, development times have to be extremely short. Design work on the project database can only start once the project science plan is available, which is usually only a matter of months before data collection commences. A major role of the database is as a data exchange vehicle and, as the total life of a Community Research Project is only five years, the system has to be on-line as soon as data are available if it is to be adopted by project scientists. The key to successful design on such short time-scales is simplicity.

Secondly, the full specification of the dataset for a complex multidisciplinary research project is unlikely to be known to the scientists, let alone the data centre, at the start of the project. Consequently, the database structure must be extremely flexible and able to adapt quickly to unanticipated changes in the dataset. Modern relational database technology is well suited to this task. The ability to change the database structure without penalty has been a major factor in the success of the project databases operated by BODC.

Thirdly, the database is to be interrogated by project scientists who are computer literate but not computing specialists. Consequently, the database either has to be designed to make retrievals of data in a form useful to scientists easy or an interface buffer has to be provided. In an ideal world, the latter approach would be taken because it avoids database design compromises away from the theoretical ideal. However, such interface development may not be possible with the technology available.

For example, in the case of the project databases operated by BODC, one design compromise was imposed to allow retrieval of data by relatively inexperienced users over a wide area network. In an ideal structure, sample property measurements would be stored in a table containing columns for:

- sample identifier;
- parameter identifier;
- parameter value;
- parameter value quality control flag.

However, scientists inevitably require cross-tabulated output to allow comparison between the different parameters measured on a sample. With modern graphical user interfaces, the implementation of such a capability from the idealized database structure is relatively straightforward. However, provision of such a capability within the constraints of line mode access imposed by mainframe computers and low bandwidth networking is virtually impossible.

Consequently, a structure had to be adopted with each parameter occupying a discrete column in the data table. This succeeded in making retrieval of cross-tabulated output trivial but there is an associated cost. Additions to the parameter set require modifications to the database structure and the associated documentation. In practice, this has proved workable but the recent migration from mainframes to workstations in both BODC and the user community should facilitate a more satisfactory solution to the problem in the future.

*Design overview*

The project database structure may be represented as a three-level hierarchy which are termed, from the top, the event description, the data linkage tables and the data tables themselves. An event is defined as an activity which results in the generation of data. In oceanography, this can range from the deployment of a sophisticated electronic package, to a water sample collected from a ship using a bucket on the end of a piece of rope. The top level of the hierarchy describes these events and positions them in terms of space and time. It therefore provides a high-level inventory of the data held in the database.

The linkage tables in the second level fulfil two functions. First, they provide the implementation mechanism for one-to-many relationships between events and data. The standard relational database technique using two key fields is used. Secondly, a linkage table is included for each major type of event. Consequently, they may be used to store information specific to that type of event. For example, the mesh size of a zooplankton net and the depth of a water sample are held in the appropriate linkage tables.

The third level of the hierarchy is the data tables. These are generally simple structures containing one column for each parameter measured, but in some cases, for example a profile down a sediment core, there is a further one-to-many relationship between the linkage table and the data.

This design satisfies the criteria set out above. It is extremely simple and therefore the detailed mapping of the specific measurements for a given project to the generalized structure may be done in a matter of days. Secondly, the structure may be easily changed. For example, an additional event type may be incorporated by simply adding a couple of tables to the schema, which is both easy to do and easy to document. Thirdly, the retrieval of any data together with appropriate header information is a simple matter for anyone who has mastered the basic SQL concept of a table join.

*Physical implementation*

BODC have implemented two project databases in support of the North Sea Project and BOFS. In both cases these were set up on an IBM 4381 mainframe running the ORACLE relational database management system. Users accessed the databases over JANET, the UK academic wide area network. The main tool used to retrieve data was native SQL supported by a small number of Pro-FORTRAN applications geared towards producing graphical images from the data.

The databases were implemented as separate ORACLE accounts. In this way, security was maintained ensuring that access to either database was limited to participants in that project. Each project scientist requiring access was registered as a user on the IBM with their own quota of disc space. This mechanism worked acceptably well in practice, but imposed a considerable administrative overhead.

The IBM mainframe ceased operation in June 1992 and has been replaced by a number of Silicon Graphics workstations on a local area network connected to Internet through JIPS (JANET Internet Protocol Service). Under Unix each database has a guest account with individual user registrations and access control managed by bespoke software. Such a system is much easier to administer.

An additional operational capability on the new system is the ability to monitor full transcripts of user sessions. This provides additional safeguards against unauthorized activities and the ability to police data usage. Two additional benefits have resulted from study of the user transcripts. First it has been possible to identify users who were struggling and offer appropriate support. Secondly, much has been learned about users' requirements and the knowledge gained has been applied to the subsequent design of CD-ROM interface software.

*Conclusions*

It can be seen that the project database is a very different entity from the corporate database that results from years of careful design work and is able to support the entire information management needs of an organization. Those used to working on the larger scale might be tempted to apply derogatory terms to the project database. However, the fact remains that project databases provide a very powerful tool in the assembly of project datasets.

BODC have used the project database concept to support two major Community Research Projects, the North Sea Project and BOFS. In both cases, over 90% of the datasets collected were assembled into the database within the life of the project. Previous data management exercises in oceanography have taken up to ten years after project completion to achieve comparable results.

## Beyond the project database

Projects supported by BODC have a limited duration and the question of the fate of the project database after the project has been completed therefore has to be addressed. As the data were to be brought into the public domain, the solution adopted by BODC has been to provide each project with a readily identified product through electronic publication of the project database on CD-ROM. The format used is a glossy package containing the CD-ROM platter, a comprehensive manual and a software interface on floppy disc.

The contents of the CD-ROM platter may be subdivided into three major categories. First there is what is termed the 'kit-form database'. A relational database consists of a series of tables which are simply a collection of rows containing a fixed set of columns. These map directly to the records and fields of a simple flat character file. Consequently, the database structure and contents may be ported from one system to another as a set of simple files, each of which holds the contents of a database table. The 'kit-form database' is precisely that.

There are many advantages to this method of data transfer. The most important of these is that simple ASCII files encoded on a CD-ROM in ISO9660 format may be read on any platform which can be connected to a CD-ROM reader. The database may therefore be ported onto virtually any system with a relational database management system. For the North Sea Project CD-ROM, a fixed format was adopted with each field occupying a fixed range of bytes within the record. Subsequent experience has shown that such a format requires greater effort to load into some systems than a delimited file and consequently a comma delimited format was adopted for the BOFS product.

The second data category is pictures. Most scientific projects generate pictorial information which may not be readily incorporated into a conventional database, and the Community Research Projects supported so far are no exception. Both CD-ROM products contain satellite images which provide truly synoptic oceanographic information which cannot be obtained in any other way. In addition, the BOFS product includes X-ray images through Kasten cores which provide information on sedimentary structures within the cores. These images are stored as bitmaps which may be viewed and manipulated using readily available public domain, shareware or commercial software.

The third category is data to support the software interface. Simple character files may be extremely portable but any application software using such simple files will inevitably run slowly as files are searched sequentially. Consequently, some of the data are replicated on the CD-ROM as compact, indexed binary files which may be accessed efficiently by the software.

The concepts behind a CD-ROM software interface require some discussion. The first aspect of such software is that unless it is extremely utilitarian and supplied as high-level language source code, there is a stark choice between targeting a specific platform or developing parallel versions. With the resources available to the BODC projects, the latter option could not be considered. Consequently, the software interfaces have been implemented for a single platform, the PC compatible. Secondly, there is little point writing software with the same capability as readily available commercial software. The interface should therefore be designed to either provide additional functionality or to smooth the interface between data on the CD-ROM and commercial software.

This may be seen by considering two aspects of the North Sea Project CD-ROM software interface. During this project, a set of predefined sites were sampled every month for 15 months. One program in the software interface allows rapid graphical examination of the time series at each of these sites. This is a project-specific requirement which could not be achieved using commercial software from the dataset supplied without several months of effort. A second program allows tables from the 'kit-form database' to be joined together and output as comma delimited files. This provides an elegant interface between the relational data and commercial spreadsheet packages.

## Future developments

The work on BOFS and the North Sea projects is now complete. However, the role of the project database within BODC is far from over. Work is currently starting on three additional project databases. Two of these support NERC projects, namely the UK contribution to the World Ocean Circulation Experiment (WOCE) and the Land Ocean Interaction Study (LOIS). The third is supporting the Ocean Margin Exchange Experiment (OMEX) under contract to the Commission of the European Community.

The migration from outdated mainframes to workstation technology with graphical user interfaces opens up exciting possibilities for future access to these databases. Client-server software provides us with the potential to build user interfaces which will make the current native SQL interface, considered so advanced less than five years ago, seem positively archaic. However, the most important point to remember is that, for the foreseeable future, project databases will allow BODC to continue to assemble project datasets of a complexity and completeness that was undreamed of so few years ago, on a timescale that is, as yet, unparalleled elsewhere in oceanographic data management.

# Transfer and SERPLO: powerful data quality control tools developed by the British Oceanographic Data Centre

ROY K. LOWRY & STEPHEN G. LOCH

*British Oceanographic Data Centre, Bidston Observatory, Birkenhead, Wirral, Merseyside L43 7RA, UK*

**Abstract:** Quality control is a vital component of data management. However, in the current data volume explosion, practising quality control with finite resources presents a very real problem. During 18 years of operation, the British Oceanographic Data Centre has developed a cost-effective quality control strategy.

In this paper we describe the quality control procedures developed with emphasis on two vital components, the transfer system and SERPLO. The transfer system provides the initial interface between external data and the BODC system, combining reformatting with extensive data checking. SERPLO is a powerful, interactive, non-destructive graphical editor. It is shown how the internal development of these systems has dramatically increased the efficiency of the data management operation.

Quality control is a major problem facing data managers in all scientific disciplines. Data are trawled from many different sources, each with their own quality standards, and are subjected to complex manipulations within the data centre to combine them into an integrated dataset. If the resulting database is to be of any value, quality control procedures are essential.

The British Oceanographic Data Centre (known as the Marine Information and Advisory Service Data Banking Section (MDBS) until 1989) has been operating in this field since 1976. This paper describes the basic elements of quality control, outlines our quality control procedures and presents two major software systems developed within the data centre which form our principal quality control tools.

## Data validation and data verification

Quality control may be considered to have two components: validation and verification. Validation is defined as the procedure by which the database administrator ensures that data submitted for inclusion to a database are not corrupted during the loading process. It is an important component in the building of a database. Indeed, external discovery of errors introduced by the data loading process will do more to destroy the credibility of a database and those maintaining it than anything else.

Verification is defined as the procedure by which it is ensured that the data loaded onto a database are correct. In some cases this may be relatively straightforward if the content of the database is restricted to material which may be readily verified externally. For example, a database containing information about organizations may be verified by contacting each organization and asking them to check their entries. Problems of scale may be encountered but the work remains firmly within the domain of clerical rather than specialist staff.

With a scientific database, such a simple approach has its limitations. One of the first problems is the definition of what is correct. This is far from straightforward. Consider a simple example. Salinity, the concentration of salt in sea water, is a basic oceanographic parameter. A chemical or biological oceanographer may regard a salinity error of 0.1 PSU to be of no consequence. However, a physical oceanographer would regard such an error as disastrous. In both cases the data value is incorrect, but in the former case the user of the data is perfectly happy to accept the data as correct.

This is a serious problem for the manager of a scientific database. Unless great care is taken, a physical oceanographer might retrieve data placed on the database by a chemical oceanographer. This is done in perfectly good faith: if one were to ask the chemical oceanographer whether the salinity supplied were accurate, the answer would be yes. The first the unwary data manager would know of the problem is when complaints arrive from a physical oceanographer who has tried using the data and realized its limitations.

A second problem with verification of scientific data is that what is correct can quite often be a matter of opinion and opinions are subject to change as scientific knowledge and understanding progress. Consider the case of black smokers. These are isolated sea-floor features

*From* Giles, J. R. A. (ed.) 1995, *Geological Data Management*, Geological Society Special Publication No 97, pp. 109–115.

discharging metalliferous waters which were discovered comparatively recently. It is not difficult to imagine a water sample taken close to a smoker being analysed and rejected as hopelessly contaminated by the oceanographers of 25 years ago.

## BODC quality control procedures

Quality control in BODC is founded on two basic principles: data examination by scientifically qualified staff and the development of stable, defensively programmed software systems. The former facilitates scientific verification within the data centre whilst the latter ensures data are not corrupted during their passage onto the data bank.

Quality control occurs at three stages during the progress of data through the system, termed transfer, screening and audit. Transfer is the first line of defence whose primary objective is the reformatting of data into the BODC internal format. However, it also plays a major role in quality control. The individual source format modules incorporate extensive data checking which trap inconsistencies in data format and errors in the data. The transfer process also standardizes the data. For example, dissolved oxygen data are supplied from different sources in units of mg/l, ml/l, μmoles/l or μmoles/kg. Transfer undertakes the necessary conversion to ensure that all data held by BODC are in μmoles/l.

The transfer source format modules are developed by professional data scientists (science graduates with programming skills) rather than by computing specialists. This does much to enhance the effectiveness of the system as a quality control tool. It is implemented using a generalized system, the transfer system described below, which enables BODC to accept data in any format without placing an unacceptable strain on data centre resources. This both encourages submission of data to BODC and eliminates the possibility of the data being corrupted through reformatting at source.

Screening is the main quality control procedure which is primarily concerned with verification but also validates transfer. It has three components:

- graphical examination and flagging of data values;
- collation and checking of header information;
- preparation of data documentation.

Graphical examination of the data encompasses a comprehensive series of checks. Spikes are identified through first difference checks. Comparative screening checks are made on contemporaneously measured parameters and on related data series. Parameter values are checked against climatological summaries to ensure that they are reasonable. Several graphical presentations of the data are used to allow them to be examined from more than one perspective. All of this work is undertaken by experienced scientists qualified to graduate or PhD level. Problem data values are marked by flagging.

Header information, such as the location where a particular measurement was made, is independently checked as it is usually supplied separately from the data, often in non-machine-readable form such as cruise reports or log books. The checks are basic but effective. For example, station positions are plotted against a coastline (it is surprising how much oceanographic data has apparently been collected on land) and instrument deployment times are checked against time channels included in the data. Much of these data have to be entered manually. Validation is achieved by keying the data twice and comparing the results or by call checking.

The data documentation includes descriptions of methodology, data processing procedures, quality control procedures, problems reported by the data originator and any concerns the data centre has about the quality of the data. It is prepared by the same staff who have undertaken the graphical examination and flagging of the data.

Data flagging and the preparation of data documentation overcome the problems with verification stated above. It is vitally important that quality control procedures do not muddy the waters for scientists of the future wishing to re-interpret 'erroneous' data in the light of modified scientific thinking. Flagging is nondestructive. The data values are therefore available for subsequent inspection and the quality status may be adjusted if scientific thinking changes. Qualifying documentation plays a major role in the prevention of low-accuracy data being used in high-accuracy applications. Users who require high-accuracy data can make an informed selection when armed with such detailed information.

Whilst quality control within BODC is nondestructive, it is common for datasets to have been subjected to destructive quality control by the data originator prior to submission. Many, including some data managers, argue that practising destructive techniques shows the courage of one's convictions. Changing such attitudes through education is a long and difficult process.

Audit is the final stage of quality control which is undertaken periodically by the database administrator on data loaded into the database. Programs are run which check the internal consistency of the data. For example, data maxima and minima are stored and the audit checks to ensure that the data lie within the specified range. Whilst this procedure is used primarily as a defence against post-load corruption, it also provides additional validation for data which have passed through the system.

These quality control procedures are underpinned by two major software systems, the transfer system and SERPLO, which have been developed within BODC. These are now examined in more detail.

## The transfer system

The concept of transfer as a combined quality control and reformatting procedure has been part of the BODC data banking model since the initial system design in 1976. In the late 1970s, large monolithic reformatting programs were developed for each format supplied. Each program incorporated thousands of lines of code and took several weeks to write. The result was a rapid build-up of a large backlog of data awaiting reformatting. It was soon realized that there was much common code between these programs and during 1981 a generalized reformatting system was designed and coded which went live early in 1982. Using this system, the backlog that had developed was soon cleared.

The system was written in FORTRAN and was initially developed on a Honeywell 66. Subsequently it was migrated onto an IBM 4381 and is now operated on UNIX workstations. Any code which has been operational for over a decade on three different platforms is stable, well tested and reliable. In the case of the transfer system, this stability has been deliberately encouraged. The core of the system has remained virtually unchanged, with over 90% of the source unmodified since it was coded in 1982. Upgrades, other than bug fixes, have been restricted to the platform migrations and have been confined to bringing input/output strategy into line with modern technology (by moving from tape to disc), replacing file-based program control with a relational database and changes enforced by differences between FORTRAN compilers.

The robustness of the system is reinforced by extensive use of defensive programming techniques. Examples of this are explicit array-bound checks incorporated into all array assignments, rigorous checks on all information passed to subroutines and frequent control checks on variable and array values. Many of these checks appear superfluous to casual inspection but the overall result is that any unexpected program action or internal memory corruption is detected and reported by the system before it can result in data corruption. These traps have rarely been sprung since the first few months of operation which forms the basis of our confidence in the system.

The transfer system is based upon a simple logical model for data which has worked extremely well in practice. This model first subdivides data into discrete physical subdivisions or 'series'. The definition of a series is one of the most important data management decisions but in most cases the subdivisions are so obvious that making the decision is trivial. The model assumes that information either has a one-to-one ('header information') or a one-to-many relationship ('datacycles') with the series. For example, a simple CTD cast measures temperature and salinity at different depths in the water column. The cast is modelled as the series, date/time and the position of the ship are modelled as header parameters and the repeated measurements of depth, temperature and salinity are modelled as datacycles.

A schematic of the transfer system is shown in Fig. 1 and a detailed description in context with other BODC systems is given in Loch (1993). The essence of the system is that a format-specific module is written for each different format handled. It is important to understand what is meant by 'different formats' in this context. A slight change in file layout such as increasing a field width from 5 bytes to 6 bytes need not necessarily mean that a new source-specific module is required. It is perfectly possible to incorporate a parsing capability into the source-specific module which handles these minor variations. Indeed, one source-specific module is so sophisticated that it will handle any combination out of a set of 20 parameters as a single 'format'.

The source-specific modules have four components:

- the channel specification table (CST);
- datacycle module;
- header module;
- trailer module.

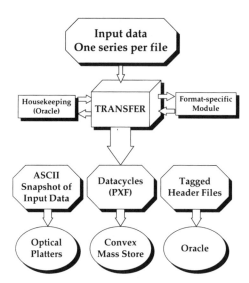

**Fig. 1.** Schematic diagram of the BODC transfer system.

The datacycles are managed by the CST and the datacycle module. The CST specifies the source and fate of each datacycle channel. A channel is sourced by specifying a pointer to either an array element or substring in a character variable returned from the datacycle module. This sourcing may be dynamic with the pointers set by the header module at execution time which significantly enhances the flexibility of the system. CST instructions can also be used to generate channels by algorithm, e.g. time channels, and apply transforms to the data values such as unit conversions. The channel fate specification controls the output and formatting of the channel to the ASCII snapshot, datacycle file or, as summary statistics, to the tagged header file (see Fig. 1).

It can be seen that the CST provides an extremely powerful and flexible mechanism for controlling the processing of the datacycles. In total contrast it is extremely easy to generate, being written in a character data description language containing only six syntactic elements. Once coded, the CST is run through a compiler which extensively checks the syntax and internal consistency of the code and translates it into a structured binary vector which drives the processing within the core of the transfer program.

The datacycle module is a high-level language subroutine which is executed once per datacycle in the series. Its primary function is to return the information from the datacycle in a binary array

and a character string as required by the CST. However, the bulk of the code in most datacycle modules is concerned with checking the information supplied in each datacycle. In some cases, such as spreadsheet output, the datacycle module extracts header information repeated on each datacyle and stores it in labelled common to await processing by the trailer module.

The header module is a second high-level language subroutine, executed once per series, which performs initialization functions and processes header information by encoding them into tagged fields. The field tags control the mapping of the information into the BODC database schema. Every check on the data which can be conceived by the programmer is incorporated into the code.

The third high-level language subroutine is the trailer module which, like the header module, is executed once per series. It has three functions. First, it encodes header information unavailable to the header module, such as the datacycle count or header information following (or contained in) the datacycles, into tagged fields. Secondly, it tidies up after series processing, closing down files and such like. Thirdly, and most importantly, it performs a series of checks on the data such as ensuring that timing information is internally consistent and that channel limits are reasonable.

It can be seen that much of the code in the source-specific modules is dedicated to automatically checking the data as they are reformatted. It is in this way that the transfer procedure provides BODC with a mechanism for automatically performing many validation and verification checks on data as soon as they enter the system.

The question which must be asked of major investments in software development is whether or not the effort expended can be justified. In the case of the transfer system, the answer is a definite yes. An experienced data scientist can develop the code for a new format in as little as a couple of hours, a reduction of between ten and a hundred fold in development time over the monolithic programs of the 1970s. Less new code is developed and consequently reliability is increased by reducing the chance of undetected bugs. All this has been obtained for less than two man-years of effort dedicated to system development.

The achievements of the transfer system in 12 years of operation are summarized graphically in Fig. 2. This shows, by year, the number of different formats handled and the number of datacycles reformatted by the system. Shown

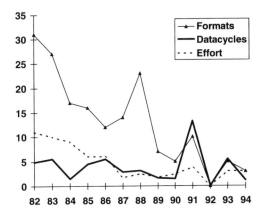

**Fig. 2.** BODC transfer system operational statistics. Formats are the number of different source formats handled during the year. Datacycles are the number (in millions) of datacycles reformatted during the year. Effort is the amount of resources (in man-months) dedicated to transfer operation during the year. Transfer was inoperative throughout 1992 due to hardware migration. Figures for 1994 are for January to March only.

against this is the number of man-months of resources dedicated to transfer. Note that this figure covers both software development to handle additional formats and operational use of the system to reformat data. In all, during 12 years of operation, 56 039 946 datacycles in 123 different source formats have been reformatted. The staff effort required to do this was less than five man-years.

This means that nearly 5400 datacycles have been extensively checked, reformatted and brought into the BODC data holdings every hour of resource dedicated to operational transfer. Such operational efficiency might be expected for high volume, consistently formatted data such as satellite data. However, in the case of the transfer system it has been achieved with diverse oceanographic data supplied in variants of 123 significantly different data formats.

## SERPLO

Graphical examination of the data supplied to BODC is an essential part of the quality control procedure. Good quality control practices dictate that every data point is examined and a flag set to describe its quality status. With increasing frequency, data are submitted to BODC which have, at best, been partially quality controlled. One has to have some sympathy for the argument that a scientist who is only interested in one or two channels from a multi-channel logging instrument should not waste time quality controlling the rest. However, there is a strong argument that those 'unwanted' channels were collected at considerable cost to NERC and consequently should be worked up to the same standard and form part of the corporate data archive. Undertaking such work has become an increasingly important role for BODC.

The problem with graphical examination is that over the last decade there has been an explosion in data volume. In 1984, the data logging facilities on RRS *Challenger* consisted of a BBC microcomputer into which data were logged by manual keying. Development work was underway to interface the RS232 port on the BBC to oceanographic instruments. Four years later, over 100 megabytes of automatically logged data were supplied to BODC from a single two-week *Challenger* cruise. NERC currently operate SeaSoar, a towed undulating platform, which has the potential for generating over a kilobyte of data per minute, 24 hours a day with a maximum cruise duration of three months.

Quality controlling large volumes of data has the potential for soaking up unacceptably large amounts of staff time unless it is done as efficiently as possible. By far the most effective way of increasing efficiency is through the use of powerful tools. For this reason BODC developed SERPLO (SERies PLOtting), a powerful, interactive graphical flag editor.

In 1986, the BODC procedure for flag editing was as follows:

- generate a hard copy plot (sometimes up to 20 m long) on a Calcomp plotter;
- examine the plot and read off the date/times of the flags to be applied;
- convert the date/times into cycle numbers using microfiche listings and enter onto a coding form;
- key the data from the coding form into an editor control file;
- run the editor against the data file;
- produce a second plot to ensure the flags had been applied correctly.

This was an extremely labour-intensive procedure requiring two screeners working together to examine a plot. It was also extremely time consuming: a plot of heavily spiked data could take several days to process. Comparative screening was a nightmare. It was not an uncommon sight to see screeners on their hands and knees in corridors examining plots that were too long to be handled elsewhere or struggling to control unruly, oversized pieces of paper on a light table. Data arrived at BODC in batches of up to several hundred series that were quickly reformatted using the transfer system. The initial plots followed a couple of days later. Shelf after shelf was being filled by boxes of plots awaiting examination. It was obvious that this manual procedure was unable to cope, even with the comparatively modest data volumes entering BODC at that time.

BODC computing at that time was based on mainframe systems: a Honeywell was in the throes of being replaced by an IBM. However, alternative computing technology in the form of graphics workstations was just becoming available and the potential offered for fast, efficient, interactive flag editing was obvious. A workstation was purchased in late 1986 and SERPLO development commenced.

The essential requirement for an interactive graphical program is speed of plotting. If the responses to user interactions are not smooth and flicker-free then fatigue will quickly set in. At the time of the initial purchase only one company, Silicon Graphics Incorporated (SGI), could offer adequate performance within budget due to their graphics-engine architecture. To take advantage of this architecture, SGI graphics primitives had to be used. Consequently, for a time, SERPLO was a platform-specific program. However, subsequent licensing of the SGI libraries has made running SERPLO on other platforms a possibility.

The program is written in FORTRAN and runs under the UNIX operating system. Between one and two man-years of effort were required for design, coding and implementation. However, in six years of operation the productivity benefit to BODC may be specified in terms of man-decades. Further, because SERPLO has made the process of thorough, effective graphical examination of data so much easier, standards of quality control have been raised.

SERPLO is both a data visualization tool and a flag editor. It is capable of displaying data in the different graphical presentations required for quality control. For example, a typical CTD (conductivity, temperature and depth) dataset includes time, depth, salinity and temperature channels. Using SERPLO, it is possible to look at such a dataset as a time series (time on the x-axis; depth, salinity and temperature on the y-axes), a depth series (salinity and temperature on the x-axes; decrementing depth on the y-axis)

or as a scatter plot of temperature against salinity. It incorporates fast, smooth zooming and panning: the viewing window can be adjusted from an hour's data to a month's data in a matter of seconds.

Intercomparison between related datasets is easily accomplished using SERPLO. One can either overlay different parameters from a single dataset or the same parameter from several datasets. The package includes a spatial plotting capability. The spatial relationship between discrete datasets, such as CTD casts, may be seen by plotting their positions on a map overlay. More powerful still is the ability to scan a channel of data collected by a ship underway with progress along the cruise track, with coastline, shown in a separate window. With such a capability data may be quality controlled in context, providing the operator has an understanding of the regional oceanographic climatology.

The manipulation of quality control flags with SERPLO is extremely easy. The data points to be flagged are enclosed by a mouse-controlled box. A single click on the middle mouse button applies the flags to the enclosed data points. A process previously requiring several minutes of two staff members' time is now achieved by a single member of staff in well under a second.

Since the introduction of SERPLO, the time taken for graphical examination of a typical current meter series (four parameters and 6000 datacycles) has been cut from approximately two man-days to significantly less than a man-hour. Even the high-volume oceanographic datasets now received containing up to 30 parameters and 100 000 datacycles per series may be graphically examined and flagged without the backlogs encountered previously developing.

SERPLO has been available to scientists at the Proudman Laboratory as part of an instrument processing system for a number of years. Recently, it has been extended to handle a range of formats, including one based on flat ASCII, and installed at the IOS Deacon Laboratory, James Rennell Centre and British Antarctic Survey. It will shortly be installed at the Plymouth Marine Laboratory. The rapid graphical examination capability provided by SERPLO is therefore benefiting an increasing number of oceanographic scientists throughout NERC.

## Conclusion

The operation of a data centre without effective quality control procedures is a pointless exercise. Pollution of a database by relatively small quantities of poor-quality data can quickly destroy its credibility as a source of useful data. Once such credibility has been lost, it is virtually impossible to regain.

However, quality control places a significant workload onto the data centre. The British Oceanographic Data Centre has developed a quality control strategy based upon a high calibre workforce supported by powerful, internally developed software tools. This has proved extremely effective in practice.

## Reference

LOCH, S. G. 1993. *An Efficient, Generalised Approach to Banking Oceanographic Data.* Proudman Oceanographic Laboratory Report No. 24.

# Data management in the National Geological Records Centre

R. C. BOWIE

*British Geological Survey, Keyworth, Nottingham, NG12 5GG, UK*

**Abstract:** The National Geological Records Centre (NGRC) has collated large collections of borehole records (600 000) manuscript geological maps (50 000) and site investigation reports (19 000) with other collections of original and archive material. The majority of these data are now digitally indexed and searches can be carried out interactively on computer or via a GIS as part of a comprehensive enquiry service.

The British Geological Survey (BGS) is a component body of the Natural Environment Research Council (NERC). One of the main terms of reference of the BGS is to 'develop and maintain a National Geosciences Database in readily accessible and usable form and an information and advisory service on geological and related matters for government, public and private sector organisations and individuals'.

The Corporate Co-ordination and Information Division (CCID) has the remit to maintain the geoscience database, to make available index level information and data from the collections and publish much of the Survey's output. Two groups within CCID provide much of the support for the geoscience database, Information Systems and Information Services.

## Organization and statutory framework

The Information Services Group is divided into three main sections.

- The Library, covering published, archive and photographic data from the UK and the rest of the world, including large collections from the previous Colonial and Overseas Geological Surveys;
- Materials collections, dealing with the storage and curation of samples and specimens, for example, borehole cores, palaeontological specimens and geochemical samples;
- The National Geological Records Centre (NGRC), which administers the BGS documentary collections of original geological data and the indexes to these collections.

NGRC is similar to the Library in that the collections are available to all bona fide researchers from industry and the academic world. However, the Library covers all published data while the NGRC covers only the UK and mainly deals with unpublished uninterpreted data.

The main NGRC collections are as follows.

- Borehole records: the collection contains some 600 000 records for the UK, including the National Well Record Collection.
- Over 30 000 geologists' field maps and notebooks: the original maps taken into the field on which a geologist has recorded his or her observations with accompanying notes.
- Geological map 'Standards' at 1 : 10 560 and 1 : 10 000 scale: these are the final interpretation of the geology after synthesis of all the available information. The collection numbers more than 20 000 maps, dating from the 1860s.
- Site investigation and road reports: over 19 000 reports donated by external organizations.
- Technical Reports: over 15 000 reports produced by BGS, with over 5000 available on open file.
- Other data include sections, information on waste sites, mine plans, photographs, etc.

The BGS has a number of statutory rights to data obtained through legislation.

- Geological Survey Act 1845: rights of access for purposes of carrying out geological surveying.
- Mining Industry Act 1926: rights to receive notification of all boreholes drilled for minerals onshore and within UK territorial waters (3 mile limit) and access to records and cores.
- Petroleum Production Act 1934: oil was defined as a mineral and therefore boreholes drilled for oil were covered by the Mining Industry Act.
- Water Act Scotland 1946 and original Water Act (England) 1945: similar rights to the Mining Industry Act for notification of and access to boreholes drilled for water over 50 feet or 15 m.
- Mines and Quarries Act 1954: rights to mine plans and access to all mine and quarry workings. These rights are equivalent to the government's Inspectors of Mines.

*From* Giles, J. R. A. (ed.) 1995, *Geological Data Management*,
Geological Society Special Publication No 97, pp. 117–125.

- Science and Technology Act 1965: vested previous acts powers with NERC.
- Water Resources Act 1991: revised the 1945 Water Act.

These acts are important not only because they give NERC/BGS rights to data, but also because we have a statutory duty to maintain those data.

## History and development

This paper will concentrate on borehole records as it is the major collection, and will also refer mainly to the holdings for England and Wales. However, the work carried out, particularly on computerization, applies to all the main collections.

The story of the development of the NGRC is, as must be the case with many large organizations, one of deliberate considered development, forced changes, committee decisions and good luck.

The Survey was formed in 1835, and in 1857 William Whitaker, considered to be the father of English hydrogeology, joined. He started collecting and registering boreholes drilled for water. In 1897 he received information from 56 wells. This was considered a very large amount of data and it was decided that memoirs should be published of all known borings in England and Wales. Publication began in 1899 with *The Water Supply of Sussex* and at that time the total collection of well records numbered about 1000.

Although diamond drilling had been introduced in the early 1870s and the first boreholes had been drilled over 1000 feet, the main reason for drilling was predominantly searching for water. However, as drilling expanded with mining, large amounts of information were being lost. Therefore, in 1926 the Mining Industry Act gave the Geological Survey statutory rights to data for boreholes drilled to a depth greater than 100 feet or 30 m.

In the following year, 1927, the Survey was notified of 108 boreholes, which were all followed up and recorded by two officers newly appointed for the task.

In 1938, well over 100 notifications were received and these now included boreholes drilled for petroleum. Over 400 water well records were received even though these were not covered by legislation until 1945. The total collection had expanded to some 20 000 records and these were registered by being numbered and plotted on New Series 1 inch to one mile topographic maps.

By the early 1950s mining had expanded to such an extent that notifications had risen to over 3000 boreholes per year and the Survey was also beginning to receive information from numerous boreholes drilled for site investigations.

The registration of boreholes on 1 inch scale maps was proving to be impractical because of the close proximity of many boreholes and the need to accurately locate and annotate their positions.

Luckily for the collections, the Ordnance Survey (OS) National Grid Reference System was introduced in the late 1950s and the system of registration on 1 inch scale maps was abandoned. The National Grid lines are published on all OS maps and allow reference to any point by a unique code comprising letters and numbers to describe the distance east followed by the distance north from a false origin. The new registration system had to be flexible enough to cope with the large numbers of boreholes drilled in urban areas and be simple enough for the field geologists who managed the collections to use and maintain. It was decided to use the new National Grid quarter sheets (5 × 5 km 1 : 10 000/1 : 10 560 scale), the new Survey mapping unit, with a simple accession number for each borehole. The combination of map sheet number and the accession number produces a unique registration number for each borehole, e.g. TQ38SW 138. A large programme was undertaken to transfer information to new site maps and re-register all the borehole holdings. These numbers are now used throughout BGS to refer to boreholes. This has proved to be a very successful and durable system which has considerably helped the development of the computer databases.

At this time, individual Land Survey field units held their own data and they were starting to be swamped by large volumes of information. The time taken to manage the data was beginning to eat into the mapping programme. Therefore, in southern England the Land Survey field units combined and set up a records unit which was a great success.

In 1973 a Borehole Records working party was set up to look at the whole area of borehole records and computerization. They stated that 'archives of borehole records are at least as important to the work of the Survey as a geological library'. At this time it was thought that there were 100 000 records for southern England, with new accessions being received at the rate of 10 000 per year. However, at this time estimates were extremely varied and often unreliable.

Work started on a pilot project to code the index and lithological data from deep hydrocarbon exploration wells.

At the same time CIRIA, the Construction Industry Research and Information Association, set up a working party to look at the case for establishing a National Registry of Ground Investigation Reports. This was reported in 1977 (Tuckwell & Sandgrove 1977) and was an important influence on the committee of Centralised Records at Keyworth (CRAK), set up by BGS to consider a central archive at its new headquarters in Keyworth. This committee considered several options, including full computerization. However, this was discarded because of cost. The favoured option was that 'Data should be held centrally or with individual units but the central index would be computerised.'

**Table 1.** Main attributes and metadata for borehole index

| Borehole attributes | Metadata |
| --- | --- |
| Registration number | Name status |
| Name | XY precision |
| Grid position | XY validation |
| Confidentiality | Data management |
| Drilled length | Earliest date type |
| Start height | Entry authority |
| Start point type | Date entered |
| Vertical depth | Date updated |
| Inclination | User entered |
| Instigator purpose | User updated |
| Earliest date known | |

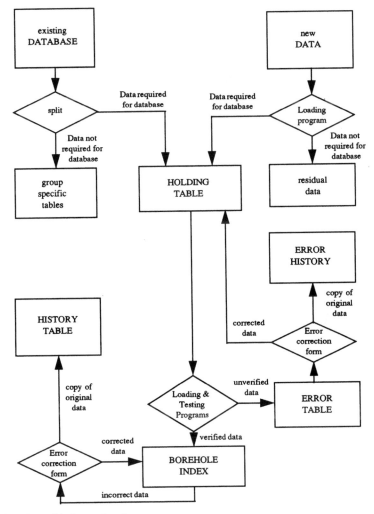

**Fig. 1.** Data flow for validation and maintenance.

R. C. BOWIE

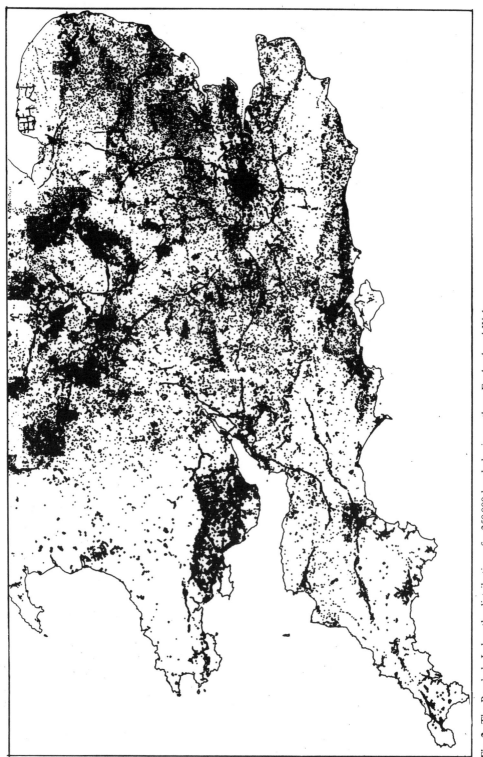

**Fig. 2.** The Borehole Index: the distribution of *c.* 250 000 boreholes in southern England and Wales.

Thus in 1984 the National Geosciences Data Centre was formed and Land Survey Records centralized (except for local borehole and geological survey data required to be held at the Regional Offices).

In 1986 the new Records Centre building was completed. The borehole records were rehoused in new archival box files in OS quarter sheet order. Standard maps, field slips and other data were brought together from various separate collections and housed in a fire-proof strong room. Improved facilities were provided for visitors to examine the data. The Keyworth Library and the Records Centre both became an Approved Place for the deposit of government records under the Public Records Act 1958.

## Development of the Borehole Index

BGS/NERC had decided to adopt the relational database system ORACLE as the organization's standard Database Management System. During 1987 and 1988, ORACLE screen forms were produced for entering index level data into ORACLE tables. Work started on all the major collections and in particular the index to boreholes. In 1990, the government decided to

provide additional funding to the 'National Geoscience Information Service' to place it in a position to be more financially secure. A scoping study of IS strategy was carried out by the CCTA and the report recommended that BGS produce a comprehensive index to all Geoscience data in the UK. The Director of BGS, Dr Peter Cook, decided that BGS should, as a first priority, have a national index of borehole data capable of being displayed graphically.

A data analysis of all the individual paper and digital borehole datasets was carried out and, after much discussion with users, the main attributes along with essential metadata for data management were finalized (Bain 1991; Table 1).

As mentioned earlier, the number of records had been poorly estimated and size of the proposed task was not known. Therefore, in 1991, an essential audit of BGS holdings was carried out so that equipment and staff could be organized (Bowie 1991). The audit showed that although many previous estimates had been very high, there were still more than 500 000 borehole records to be processed. A group was set up to carry out the work and a data manager appointed. This was essential to keep track of all the data.

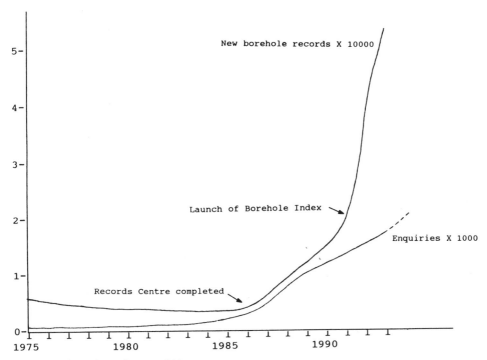

**Fig. 3.** Enquiries and new data acquisitions.

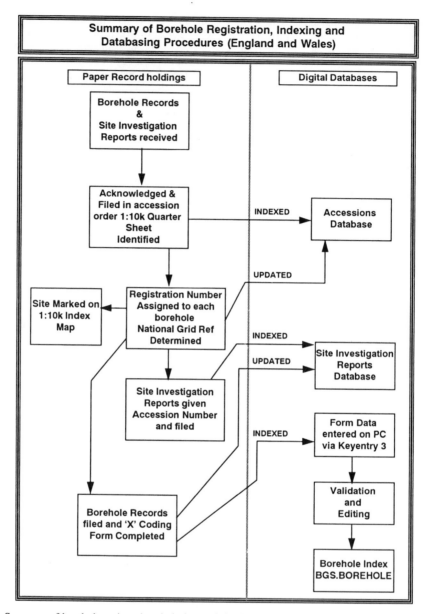

**Fig. 4.** Summary of borehole registration, indexing and databasing procedures (England and Wales).

Data were gathered from existing digital indexes, paper registers or coding forms and maps (Fig. 1) (Smith *et al.* 1993). It is important to emphasize that no new coding was carried out as it would have been too expensive in time and resources. The index level data were either entered directly onto ORACLE tables using ORACLE forms on VAX mainframe terminals or onto PCs using the data entry program KeyEntry3. Positional data were also recorded using a digitizing table and the graphics package AutoCAD. The data were then run through loading and checking programs to amalgamate the data and eliminate major errors (Laxton 1989). Some verification was carried out by comparing the positions determined from digitized maps with those from coded entries. However, in the end the principal method of verification will have to be by use. Spot checks have shown that errors in data entry are less than 1% and, importantly, insignificant for the grid position.

The Borehole Index was officially launched in 1992 and the current position in southern England and Wales is shown in Fig 2. The distribution of boreholes mainly reflects development in major urban areas, road building and mining activity.

At the same time, work continued on the entering of index level information to ORACLE tables for the other collections.

## Present position

Due to the improved access to the holdings and because of the increased awareness of the databases, the NGRC are receiving increasing numbers of site investigation reports and borehole records. As can be seen, the number of boreholes received has gone up nearly 20-fold since the mid-1970s. The number of enquiries has increased 200-fold in the same period (Fig. 3).

Current procedures for coping with this influx of data are shown in Fig. 4. These are part of a documented quality management process. The main points are that all new data are databased and all boreholes are given individual registration numbers. The paper holdings are shown to the left and the digital databases to the right.

All new data are located on a 1 : 10 000 scale quarter sheet, acknowledged, databased with basic information and filed awaiting registration. This allows the NGRC to monitor what data is being received, who it is from, whether it is commercial–in-confidence, the number of boreholes in the reports and whether they have been registered. Site investigation reports are held separately because of their bulk and because the copyright to some of the information they contain may belong to the client. Each borehole within a report will, however, be registered and a copy of the record held in the main borehole collection.

## Access to information

Access to the index-level digital data is by several means:

- by direct interrogation of ORACLE forms;
- by interrogation of the ORACLE tables by SQL (sequential query language);
- by means of a menu-aided retrieval system MARS designed in-house. This provides an easy-to-use menu system which is easily updated and has the facility for custom or personal menus to be incorporated.

MARS is a combination of a PRO FORTRAN program and a purpose-built ORACLE database table which together act as an interface between the user and computer system. This shell allows data to be moved between program modules with the minimum of involvement from the user, giving the impression of a single system. The user is also isolated from the complexities of the ORACLE SQL program language (Robson & Adlam 1990).

Finally, the NGRC has a sophisticated Geographic Information System, the BGS Geoscience Index (Fig. 5). This system is implemented on Sun workstations and uses Sun IPC workstations connected by local area network to Sun 630 database servers. Sun IPX stand-alone systems are used at the regional offices and information points. The system is designed for any geographically referenced data where the links between datasets are spatial and searches are for specific areas (Loudon et al. 1993).

The primary purpose of the Geoscience Index is to improve access to BGS data. The GIS system used is the commercially available ARC/INFO linked to ORACLE DBMS with a user-friendly front end designed by the BGS Information Systems Group and written in ARC MACRO LANGUAGE. The system allows for virtually any spatially related data to be added and there are now a large number of datasets available. The system is almost completely mouse-driven and the remit was that the instructions should be simple and the system should be easy to use and suitable for occasional or unskilled users. At present, access is only available to BGS staff through the appointed workstation managers.

The system allows for an area to be chosen by various methods including the Ordnance Survey (OS) gazetteer of 250 000 place names. The 'backcloth' is by default a scanned OS topographic base, the scale of which is dependent on the area selected. In addition, vector data, e.g. administrative boundaries and National Parks, etc., can be displayed. The geographic spatial reference units used can be varied and data from more than 40 datasets chosen and displayed.

Selected queries including complex SQL programs can be carried out on screen. The results of these queries can be produced as lists and plots at local or remote printers. Although at present these searches only provide index-level information, the system has the ability to display scanned images of the documents associated with particular locations. Work has started on scanning the paper borehole records.

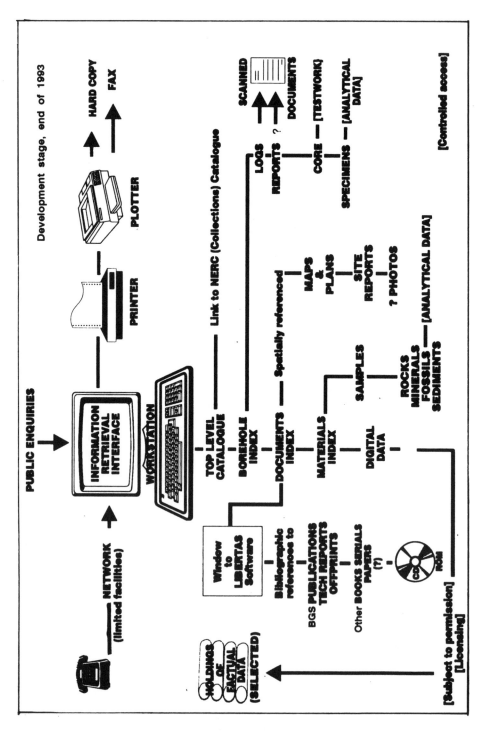

**Fig. 5.** The NGRC Geographic Information System.

In certain areas we can display vector data from the 1 : 10 000 scale geological maps produced by the BGS digital mapping programme. Alternatively, there is 1:250,000 scale raster scanned geological data.

Prior to release, all new versions are tried and tested by users within NGRC. External users of NGRC and BGS Information Services have also been surveyed to determine their requirements. The system is currently on version 4 and development work is in progress on releasing a CD-ROM version.

## Summary

The NGRC has brought together a large quantity of geological data. It is now all being digitally indexed. Much more is now known about the data held. It can be displayed graphically or provided as lists or plots as part of a comprehensive enquiry service.

## References

BAIN, K. A. 1991. *Data Analysis for a BGS Borehole Index*. British Geological Survey Technical Report **WO/91/2**.

BOWIE, R. C. 1991. *Report on the Borehole Records held at the BGS Offices in England and Wales*. British Geological Survey Technical Report **WO/91/6**.

LAXTON, J. L. 1989. *Documentation of the Oracle borehole database data loading and checking system*. British Geological Survey Technical Report **WO/89/9**.

LOUDON, T. V., ADLAM, K. A. M. & GIBSON, J. R. 1993. *Geoscience Index System (Version 3.0) User Guide, 1993*. British Geological Survey Technical Report **WO/93/11**.

ROBSON, P. G. & ADLAM, K. A. M. 1990. *The Mars User Guide (Menu-Aided-Retrieval-System)*. British Geological Survey Technical Report **WO/90/1**.

SMITH, I. F., ADLAM, K. A. McL. & BRIERS, J. H. 1993. *The British Geological Survey Borehole Index Database: A User Guide*. British Geological Survey Technical Report **WK/93/24**.

TUCKWELL, D. J. & SANDGROVE, B. M. 1977. *A case for a National Registry of Ground Investigation Reports*, CIRIA.

# Geoscience data value, cost and management in the oil industry

RICHARD G. MILLER & JOHN S. GARDNER

*BP Exploration Operating Company Limited, Uxbridge One, Harefield Road,,
Uxbridge, Middlesex UB8 1PD, UK*

**Abstract:** BP Exploration, through its Frontier and International division (XFI), is the first oil major to openly release some of its international exploration data, information and internal reports to competitors for cash. The data and information relate to geographical areas in which the company foresees no strategic interest.

There are various ways in which such a programme could be run. In our case, all exploration information and data have an identified owner, usually the asset manager responsible for exploration in the relevant part of the world. A dedicated XFI data broking team identifies and prices material for possible exchange or licence sale. The asset manager who owns the material certifies whether it may be released without damage to the company's competitive position. Legal staff check the ownership of the material, and ensure that there are no confidentiality constraints. Commercial brokers then either offer released material to the open market or create new products from it for licencing. The material is licenced for use rather than sold, so that the recipient cannot re-sell it. All income accrues centrally to XFI rather than to the asset manager; altruism by the owners of the data is central to our data brokerage philosophy.

Such marketing follows naturally from treating data and information as a true asset, whose value can be established in the market-place. When such marketing becomes the industry standard, then XFI can also enter the market-place as a buyer. Those who can buy existing exploration data and information will save the great cost of generating their own. This is part of a sea-change in the industry, in which the players are moving towards a lower cost base through shared activity as opposed to duplicated activity.

Several years ago BP Exploration's Frontier and International Division, or XFI, embarked upon radical changes in the way it maintained its exploration data and information. In the process it became clear that the company owned extensive information resources which it no longer needed. This information was a frozen asset. It was expensive to maintain, and clearly of some value, yet of no direct use to the company. XFI wanted to realize this value. It was also seen that XFI could radically cut the cost of acquiring new information and data if it could buy what already existed, rather than re-generating it.

For some years seismic data have been freely traded, sold or licenced within the United States and a few other countries. Various types of international data have also been traded (exchanged) between companies. However, the idea of releasing general international exploration data and information for cash was essentially new to the industry. XFI therefore set up a two-man data brokerage team in 1991 to explore ways of releasing financial value, as income or savings, from unwanted data and information. The task was broken into distinct steps:

- the releasable information or data which are not germane to any part of BP Exploration had to be identified;
- the ownership and any confidentiality constraints had to be determined;
- a market had to be identified;
- a marketing method had to be settled, both for sales and for trading;
- a realistic view of the value of each item, both to purchasers (for setting prices) and to BP Exploration as a whole, had to be acquired.

In practice, the data brokerage team first selects potentially releasable material, usually at the request of a potential customer. There follows a legal check, an overview opinion from an internal BP Exploration consultant, and sanction from an identified owner (an exploration asset manager). Retained brokers are then authorized to market licences for released material.

The distinction between licencing and selling is crucial. A licence only allows the buyer to use the information for his own exploration purposes. He has no ownership, he cannot sell the information on to a third party, and he must keep the information confidential. A few licences permit a contractor to undertake some form of re-processing or other *value-addition*; the new product is then licenced under specified profit-sharing terms.

*From* Giles, J. R. A. (ed.) 1995, *Geological Data Management*,
Geological Society Special Publication No 97,pp. 127–135.

The project has now proven its worth. In 1993, XFI sold licences worth US$1 million to use selected data and information. All costs have been recovered, and XFI now has a new stream of income. However, direct income from licence sales is not the only goal. The sale and exchange of non-sensitive information will probably become accepted practice among major oil exploration and production companies. When that happens, considerable savings could be made by acquiring information in the open market and avoiding the cost of generating such information. XFI see no benefit in re-inventing wheels.

This paper describes why XFI has adopted this radical but rational view. It then describes how the project is now working, and the details of the brokerage and trading methods that have been selected.

## Data, know-how and information

### Definitions

There are no commonly accepted definitions of data and information in use within the oil industry. Except where noted otherwise, the term 'data' here refers to simple facts, usually measurements (such as seismic data), or 'pseudofacts', usually unequivocal observations (such as lithology). These data are usually unusable without some level of skill, understanding or know-how which produces an interpretation. There is a somewhat grey area between data and interpretations, in particular where measurements depend upon the particular technique, equipment or technician involved. For example, some geochemical measurements are equipment-dependent, and palaeontological data depend entirely upon the skill and knowledge of the observer.

Data and know-how together create an interpretation, knowledge or information. This can be expressed as:

$$\text{Data} \times \text{know-how} = \text{information (or knowledge, interpretation, etc.)}$$

In this paper we prefer the term 'information' because we are literally informed and can make better decisions.

Contrary to popular belief, information alone is not power but only a potential power, a possible technical or commercial edge. It has to be used, and used properly, in order to profit. Data, know-how and information could all be freely released if their owner could be sure of using them better and sooner than any competitor. However, early in the data brokerage project, it was decided that only data and information would normally be released to third parties. With rare exceptions, notably certain in-house training courses, the know-how and technical skill of BP Exploration staff will remain confidential. This can sometimes be a fine distinction to make, but it helps to protect against undue loss of competitive position.

### Maintaining industrial data and information

Tangible data and information in the oil industry consists of hard copy and electronically stored data. It includes books, reports, maps, seismic sections, photographs, digital data of all types and databases. A lot more data and information is intangible because it is stored in the medium of staff intellect, but obviously this is very difficult to treat rigorously. First, it is hard to catalogue this knowledge and make it generally accessible without a very wide, informal network of personal contacts. Secondly, it is difficult to define either its cost or its value. Thirdly, it is impossible to be sure that the memory is reliable and uncoloured by personal inclination.

A company's tangible information has distinct users, custodians (curators) and funders, roles which are sometimes combined. The principal users are typically the technical staff. The custodians vary between companies. XFI has a formal, centralized system of storing data and information. Custody of most of the physical materials has been outsourced, that is, passed to external specialist contractors. The custodians, both within the contractor companies and those remaining within XFI, are all specialists. In other organizations custody may be the responsibility of other staff, such as an IT group, or of the data gatherers and users themselves. The cost may be borne centrally, or individual teams may bear the cost of their own data, or ultimately each user may fund the full cost on a 'user pay' basis.

A centralized information system with centralized funding can be extremely effective at storing and retrieving material, but at a relatively high cost. In such cases the stereotypical user tends to abuse the service because it is free at the point of demand, while the stereotypical custodian is a user-hostile collector of all information regardless of value or cost. This situation has led to the paradox of the 'low-value crown jewels'. A company's holdings of data and information frequently have no formal book value, which makes it difficult to justify the cost of its upkeep, yet the collection is simultaneously regarded as being too valuable to lose or to release outside the company. A service organized in this way has too many inherent paradoxes to function efficiently. The aims of the users, the custodians and the funders are not aligned.

All XFI data and information has a specific owner, generally the asset manager for the relevant geographical region of the world. The whole collection is available to all. Physical curation by the external specialist service providers is managed by a small data service team. Only a few peripheral services, such as management, directly requested by asset managers and supplied by BP Exploration staff, are directly and jointly funded by the asset managers. Otherwise, all users, including the data owners, pay the service provider for services on a per-item basis, for example for storage, retrievals or data loading. This turns users into user-funders, forcing them to recognize the cost every time they use the service, and to assess whether the benefit is worth the cost. Our information resource is entirely demand-driven and funded by its users, needing no central financial support.

## The value of data and information

### What is value?

Part of the philosophy behind a more rational handling of XFI's data and information was to realize that this resource is not a burden but an asset. True assets need to have their book value established. Value rarely equals the original cost. Value is what someone would pay now to own the commodity. An old seismic survey may originally have cost millions to shoot, but it might not command a fraction of that price on the open market, even if sold several times over. Conversely a perceptive and original analysis of the same data may have cost very little in man-hours, but if it can prevent the drilling of a dry hole its value may be many times its cost. We know of no oil company that has ever either valued or costed its data and information.

### Establishing external value

The external value of data and information can only be established by the market. In general, when an asset is worth more to a competitor than to its owner, then it should be sold. This is the step which XFI is now taking. Because of XFI's avowed long-term exploration strategy, parts of the information and data collection are now of little direct value to us, but of considerable value to others. This is the material which we have started to identify and release on to the open market. Licencing data and information unequivocally defines its value. This value should prove that the information is worth proper curation. The income could even be used to offset curation costs, although XFI has not chosen this route.

### Establishing internal value

It is relatively easy to find the external market value of unrequired data and information, but much harder to set its internal value to the company. This value might seem very low until we consider how the company's competitive position might change if a competitor acquired the material. There seem to be four broad themes in discussions about the effect on competitive position.

First, it can be argued that the competitor who acquires this information lowers his exploration costs, and therefore competes more effectively with XFI on a financial level. However, an exploration budget is often fixed beforehand and tends to be completely spent, so the principal financial change is that some of that budget is paid to XFI rather than to a contractor.

Secondly, it can be argued that acquiring BP Exploration data and information may prevent competitors from wasting money on major exploration activities, such as seismic surveying or drilling, over poor acreage. However, it seems unlikely that third-party data alone would be sufficient to tip that balance of decision. Critical data would in any case be priced accordingly.

Thirdly, competitors will be enabled to identify the same prime exploration acreage as XFI, and to compete for that acreage with the same technical case. Effectively all competitors are brought up to the same level of expertise and ability if all information is shared. However, different explorers will always view and use the same information differently. Furthermore, XFI will not release information from areas where we intend to explore aggressively.

Finally, all these possible benefits to competitors are precisely those which XFI would like to acquire by establishing a market which we can enter as a buyer. All else being equal, we then lose as much by selling as we gain by buying, so how do we benefit? We believe that we can nevertheless sustain our edge in our areas of interest. Today's severe cost pressures are forcing the exploration industry to cut exploration costs, and one way to do this is to move to a shared knowledge base. The greater advantage will go to those who make the transition to sharing information first. Eventually, the critical advantage will not come from possessing information, but from using that information better and faster. Future competition will occur in a different conceptual arena.

## Selling and buying information

The rationale behind buying information and data is that it frees professional staff to think about it and to make decisions with it, instead of having to generate it. Probably the greatest initial work-load on explorers is the task of acquiring information. As an industry we have undeniably re-invented a lot of wheels. Geoscientists continually assess and re-assess the potential of exploration acreage and prospects, but to a great extent they are only repeating the work of their peers in other companies.

Selling (or licencing) one's data and information is a sensitive topic. At the extremes, it can be likened to selling the family silver, while prospective buyers may suspect that only relatively worthless material would be offered for licencing. These are natural suspicions in a new and untried market, but neither view is correct. Such attitudes conflict with the fact that the oil exploration industry supports many consultants and companies who create and sell data and information. Another company's information is generally no worse, or better, than your own, but it is usually different. Yours is not the family silver, and theirs is not rubbish. A company comfortable with the logic of releasing information should also be comfortable with buying it from competitors. When selling information, it may be some comfort that, like few other products, you still possess information even when you have sold it. With regard to buying information, 'made in-house' is not a criterion for quality, only for familiarity and comfort.

## Data brokerage

'Data' brokerage as used here loosely includes all types of data, information and other exploration-related materials. 'Brokerage' refers to the use of professional brokers to handle marketing and licence sales.

There are various ways in which data and information could be offered for licencing. The process adopted by XFI is not necessarily ideal for all companies in all circumstances, but it fits our particular philosophy of data ownership and care.

## Why brokerage?

XFI have considered various ways of releasing information into the market-place. For example:

- XFI could employ brokers to market materials and sell licences;
- XFI could sell licences itself;
- data could be traded or released for credits in a data-exchange run by a broker (i.e. cashless dealing apart from the broker's commission);
- a group of companies could found a mutual data-exchange.

BP Exploration information is owned by different assets, so the data-exchange options would require a 'seller' asset group to simultaneously be a 'buyer' too. In such cases, the asset manager is unlikely to need help from the data brokerage team. More commonly, the seller is a dormant asset with no interest in acquiring data. BP credits from sales could, of course, be pooled for the active assets to use for purchases, but some higher authority would then have to allocate the pool. If sales and purchases did not match in value, then cash dealing would still be required to reach a balance. Valuation of materials in the exchange would probably have to be in money terms anyway.

XFI therefore chose a cash-based system as the simplest option, using external brokers because BP Exploration is not a professional brokerage company. Such companies already exist, and we should not try to compete in their business. The philosophy of outsourcing suggests that a task should end up with a company for whom that task is core business.

Cash-based professional brokerage still permits data trading (exchanging). A good broker can organize simultaneous trades between several companies, exchanging different types of materials from different countries. In these circumstances only a neutral intermediary stands much chance of reaching general agreement on equivalent trade value.

## XFI data broking principles

*(a) Centralisation.* XFI data and information belongs to specific asset managers who are free

to manage their own data and to trade, licence or release it as they choose. However, in practice almost all licencing and some trading is organized by the centralized data brokerage team. This streamlines the process, giving efficiency and a single point of contact and expertise in all matters of data release or purchase. It is also a necessary adjunct to altruism.

*(b) Altruism.* For several reasons, the proceeds from licencing are credited to XFI in general, rather than to the data owning assets. An asset's exploration budget is set by BP Exploration and should not be artificially bolstered by extra income from licencing data and information. It could be counter-productive to XFI as a whole if material was released independently by asset managers, simply to satisfy local budgetary demands. Releasing data should not jeopardize any other asset. Some assets rather arbitrarily have a large endowment of releasable information, while others have none. Altruism effectively maximizes the overall benefit to BP Exploration.

Altruism also resolves some problems of data trading. Trading can be easily handled by the relevant asset manager where the incoming and outgoing data relate to the same part of the world. In other cases, however, XFI wants to exchange data from two different countries, one where we have a long-term interest and one where we do not. This involves two different BP Exploration asset managers. The owner of the dormant asset must give away the data traded out, for the benefit of the active asset which needs the data traded in. Without altruism, data owners might prefer to sell licences and keep the proceeds; active assets, which have the greatest need for information, would have nothing to exchange for it. The data brokerage team actively searches out data from competitors when requested by an asset manager, and trades data from other assets for them. The data-receiving asset only pays for the man-hours of the data brokerage team; the traded BP Exploration data are donated by their owner.

Altruism is helpful in value-addition projects where information from several assets is combined and re-packaged for licencing, for example as a single regional report. It is preferable that all relevant owners should release the required materials freely and that proceeds should be credited centrally.

Centralization and altruism combine to maximize the benefit to BP Exploration as a whole. Nevertheless, they are only guiding principles.

The asset managers always have ultimate control over their own data and information. The data brokerage team can only act with their express agreement.

## The brokerage process

*(a) Choosing brokers.* Brokers should be chosen by a normal process of competition and tendering. However, if more than one broker is retained, they should not then compete with each other, otherwise customers can play them off against each other to achieve lower prices. This ultimately damages both the brokers and the company releasing the information. XFI's different brokers have different fields of activity, for example simple brokerage and the creation of value-added products. In practice the brokers have exclusive rights, but this can be altered by BP Exploration as circumstances dictate.

*(b) Releasing and licencing.* The licencing of reports is the most complicated form of brokerage, and it illustrates all aspects of the process (Fig. 1). BP Exploration has a culture of writing formal reports on all exploration-related work. The XFI collection numbers around 100 000 items, from regional reviews to single-sample analyses.

XFI's release process starts with the identification of geographical areas where no strategic interest is foreseen. This does not mean that BP Exploration sees no petroleum potential in these areas, rather that exploration strategy is already mapped out, targets selected and budgets committed. Barring the unforeseen, we believe that we are already best placed for exploration success. Legal staff produce a summary of all legal agreements and legislation affecting data and information that BP may have from these areas.

The release process then moves to three 'levels'. At 'level 1', if the asset manager agrees, the basic details of all reports pertaining to the releasable areas, such as title, author and date, are printed from the catalogue and passed to the external brokers to show to prospective customers. At this stage, none of the reports has been checked for releasability.

At 'level 2', a customer has requested more information. The report is retrieved from the collection, and the contents list and the abstract, edited as necessary, are copied and sent to the broker for the customer. At this point, if it is already evident that a report is not releasable, it is withdrawn.

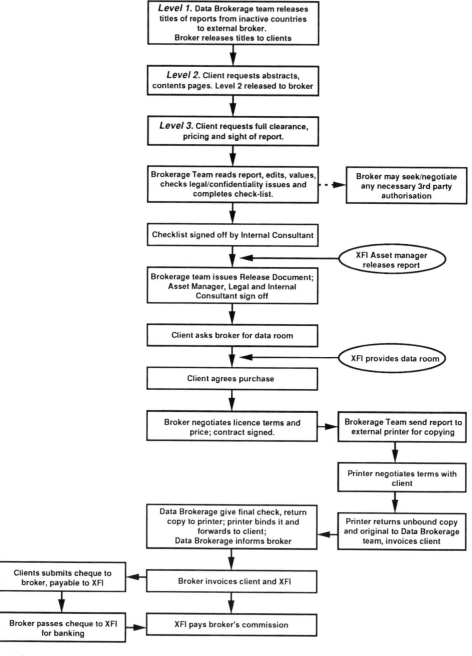

**Fig. 1.** Report brokerage in XFI-BP Exploration.

At 'level 3', the data brokerage team reads the report. They complete a check-list of releasability criteria, such as third-party ownership, other permission requirements, risk to competitiveness or reputation, and government restrictions. The check-list is signed by an internal BP Exploration consultant who can view the report in its broad context to the company. The asset manager is asked for permission to proceed. If the report is releasable, a price guide is set. Criteria for pricing include the report's age, the price of equivalent contractor's reports, the quality, quantity and rarity of the data included, and the level of industry interest in the geographical area concerned. A release document is issued, listing the price guides and any conditions attaching to the release (such as third-party permission). This document is signed by the consultant, the legal department and the asset manager. It is then passed to the broker, whose authority it is to market the report. It can be withdrawn at any time.

Customers must be made aware from the outset that a level 1 or level 2 release does not necessarily mean that a report will ultimately be releasable; we must manage their expectations. It is an unfortunate drawback to the process, but it is not practicable to read every report before testing for customer interest. The average reading load is six reports every working day. About 1500 reports have been checked at level 3. It is felt that customers know their own needs best; they should decide which reports are of most interest and worth checking for releasability.

After level 3 clearance, customers may briefly view chosen reports on our premises, as a quality assurance. Any further negotiations then continue between the customer and the broker, who has authority to negotiate prices within reasonable bounds, for example in constructing package deals and setting discounts. When an order is placed, the broker negotiates a licence agreement within bounds set by XFI. This licence forbids the buyer from publishing or passing the information on to other parties, and specifies the licence fee and method of payment. The data brokerage team forwards the report originals to a commercial copying agency. The cost of copying is arranged directly between the customer and the agency, and is payable by the customer. Before binding, the copies are briefly returned to the data brokerage team for a last check. After binding, the copies are despatched by the agency directly to the customer.

This general process can be adapted for licencing other forms of data, information and exploration materials. A simpler process covers value-adding projects where information is incorporated into a single commercial product. Two guiding principles underlie the whole brokerage process for our security. First, all materials always remain under XFI's direct control except during the copying stage. Secondly, all licence fees are payable directly to XFI, which pays commission fees and any income due to third parties. Within these constraints, as much of the process as possible is handled by the broker.

## Ownership, governments and third-party interests

Ownership of data and information is a critical issue. Since exploration is commonly a joint enterprise, several partners may have joint rights to the data and information produced. The idea of releasing information afterwards for cash is relatively new, so contracts do not always specify who owns the data and information after the contract has expired. Even when data ownership is defined, the ownership of interpretations can still be uncertain, and the boundary between data and interpretation can be left indeterminate.

Partners who are joint owners of data and information must give their agreement to any sale of licences. We have found that they will generally give the necessary permission in exchange for an income share (often taken in the form of XFI information). The greatest difficulty is the long delay involved in finding and contacting staff with the necessary authority to make a decision. Obtaining third-party consent is sometimes a delicate matter. In some cases this task can be passed to the brokers, but not when dealing with national authorities.

Governments and National Oil Companies (NOCs) have unique ownership concerns. Governments often set exploration licence terms which include government ownership or co-ownership of all original exploration data obtained on their territory, principally seismic surveys and electric well logs. This right of ownership sometimes extends to all information, including opinion and interpretation, which is obtained or created as a consequence of the exploration licence. In such cases the government has the sole right to licence their material. Some governments, however, which are not equipped to supply such data to the industry routinely, effectively and efficiently, have allowed XFI to licence its data on an income-share basis.

Other governments prefer that no data or information should be released, perhaps in case it reduces the attractiveness of their acreage. We have now encountered several companies that will no longer consider bidding for certain acreage without 'neutral' information which has not been selected for release by a national authority. Perhaps a country's interests are best served by an open policy regarding data and information, however its release may be organized.

Very commonly a report incorporates material derived from other reports which are not owned by BP Exploration. It can then be helpful to consider the 'Deconvolution Principle' when deciding whether the BP Exploration report is a genuinely new interpretation which we therefore own outright. Could someone with the BP Exploration report deconvolve it to obtain something close to the original material? If not, then a new product has been created. For example, digitizing a map does not create a new product, but processing satellite data into a photographic image does. A related issue concerns the ownership of analyses performed on traded rock or fluid samples.

## Progress

### Brokerage income

To date, the data brokerage project has run for over two years. The level of business reached US$1 million in 1993. A lot of time was initially spent obtaining legal opinion on the status of data and information ownership from various parts of the world. The first thrust was towards licencing hard data, such as seismic and electric well logs, but eventually it became clear that BP Exploration rarely had outright ownership of it. We have now focussed more upon licencing interpretative reports. Our success in selling these reflects well upon the quality of BP Exploration work. A quality report can convince a client of the merits (or otherwise) of specific exploration activity, and encourage them to acquire the relevant data (or not).

### Value-added projects

Value-added projects were started in 1993, and have yet to reach their full potential. Most such arrangements give XFI royalties and a copy of the final product. Among the ongoing projects are:

- supplying a database to a consultant company, for merging with their own version; XFI received the merged version and regular up-dates;
- supplying consultants with general exploration information on acreage involved in upcoming licencing rounds; this is incorporated in reviews of the exploration potential for the general industry;
- supplying core, cuttings and oil samples for analysis; the analyses are then made commercially available;
- supplying old seismic for re-processing and commercial sale;
- passing non-sensitive, in-house, training courses to a commercial training agency; XFI receives royalties and future training at cost, and avoids the need to update and run future courses.

## The future

### Development of the market

As noted before, XFI intends to establish a general industry market in data and information. Some other exploration companies are now preparing to broker data. Frequently this is driven by a lack of funds for the purchase of XFI data. The growth of this market is commercially logical and probably inevitable. There are not many other ways to trim exploration budgets further. This market provides a mechanism either for generating cash from frozen assets or for obtaining exploration information at lower cost. The exploration companies which will thrive best may be those that extract the most value from all their assets, including their data and information.

We cannot predict how the contractor market will adapt to this initiative. If exploration budgets are fixed or diminishing, then money spent with BP Exploration is presumably taken directly from this market. It seems likely that future contractor surveys and research will more often be commissioned rather than speculative, as costs and risks rise. It may be much cheaper, equally effective and less risky to update data and information that already exist. Perhaps the contractors who will thrive will be those who reach agreements with data-rich exploration companies.

As trading, licencing and accessing data and information become more important, future joint exploration contracts should specify more clearly the ownership of information acquired jointly during exploration. The ambiguity of this question is presently responsible for many delays in releasing data and information for licencing.

If a company recognizes that it can defray the cost of a project by selling the data at some future date, then the work could be carried out with this in mind. For example, any third-party data which were used could be clearly identified for easy editing. A contractor might also charge less for a commissioned project if he is offered future selling rights. However, the aims of the contractor would then no longer be exactly those of the company, which might pose a slight risk.

If the data market does grow as expected, then XFI will soon have significant competition in this field. This will bring down the price of data and information, as more sellers appear. By then, however, XFI may have licenced its best material at the best price, and could enter the market as a buyer as prices fall. In any event, XFI's income from direct licencing will probably eventually fall. We believe that the longer-term market is in data integration and value-added products, where we should reap increasing benefits in cash and kind from close co-operation with commercial information organizations.

*Specific issues*

Entering into data brokerage has identified a number of uncertainties in the field of technical intellectual property as it applies to the petroleum exploration industry. Some examples are listed below.

- Can we define data, information, know-how, 'exploration materials' etc.?
- What is accepted practice?
- If one company gives another a sample which is then analysed, who has what rights?
- Does the industry already have most of the necessary data to tap the world's YTF?

- What is the value of a large company's data holdings to that company, and what is the value to a third party?
- What would it cost if you lost it, particularly into the wrong hands? What is loss prevention worth?
- What is the copyright law on imagery, maps, text and electronic data?
- Are the successful companies the most secretive or the most open?

## Concluding remarks

XFI has taken a controversial step in opening up part of its data and information resource to competitors. It has been necessary to create a process for doing this, because it has not been tried before. Nevertheless, it is believed that it is part of the natural evolution of the business. If the idea takes root and establishes a general market, then XFI will in future have the privilege other companies now have, i.e. the ability to obtain another company's perspective on an exploration topic. Although licencing information has been profitable, the greatest benefit may lie in becoming a buyer.

It has been interesting talking to visitors who have come to see what is on offer. We have been encouraged by the response to our initiative so far. Those who believe we have taken a bold and rational decision outnumber those who are deeply suspicious of our motives. It is the authors' personal belief that the exploration companies cannot afford to fight each other down to the last survivor; the last survivor, after all, may be someone else. In the present economic climate, despite the recent (and perhaps very temporary) rise in the oil price, we could all use more co-operation.

We thank BP Exploration for permission to publish this paper, and the Geological Information Group for providing the opportunity. Some views expressed are those of the authors and not of BP Exploration; we hope that the text makes the distinction clear. Finally, we thank the old sweats within Data Services who built the new philosophy from which data brokerage arose.

# Groundwater level archive for England and Wales

P. DOORGAKANT

*British Geological Survey, Maclean Building, Crowmarsh Gifford,
Wallingford, Oxon OX10 8BB,UK*

**Abstract:** The monitoring of groundwater levels is important for the management of groundwater resources. Within the United Kingdom a selection of observation boreholes are monitored for periodic assessment of the groundwater resources. The measurement of groundwater levels is undertaken principally by the water companies, the National Rivers Authority (NRA) for England and Wales, the River Purification Boards (RPB) for Scotland and the Department of the Environment (DoE) for Northern Ireland. Weekly and monthly data are supplied to the British Geological Survey (BGS) and these data are entered into an ORACLE database where they are assimulated with historical records. Additionally, paper copies are archived. Data usage is facilitated by a number of computer retrievals including well hydrographs in graphic form. Access to the groundwater level information is made available to the public at a small operational cost. A number of publications are produced giving a review on groundwater, including an assessment of recharge to major aquifers and also a number of hydrographs are illustrated from the key boreholes. The database has proved to be an invaluable tool in predicting environmental change, such as droughts and rising groundwater levels.

Groundwater has been an important source of water supply for many centuries and now constitutes about 35% of the supply of England and Wales. The reliability of groundwater supplies and the contribution of groundwater baseflow to river flows and the aquatic environment was thrown into sharp focus during the 1988–1992 drought. The drought severely affected the English lowlands and by the summer of 1992 groundwater storage throughout England and Wales was at its lowest this century (Marsh *et al.* 1994).

The maintenance of a national groundwater level archive is a central government responsibility. Since 1982 the archive and the associated publications have been maintained and prepared by the Institute of Hydrology and the British Geological Survey (BGS) at Wallingford on behalf of the Department of the Environment (DoE).

Archive usage has increased significantly in recent years; there is now a regular monthly nationwide overview of the groundwater situation. The archive is used during the preparation of yearbooks, reports, journals, etc., and for media briefings especially during periods of extreme climatic events (flood, drought, etc.). Additional benefits include the provision and access of data to users and exchange of ideas with measuring authorities, resulting in further development, output and the safekeeping of data.

The groundwater archive is an important asset for understanding and improving the management of groundwater resources. The archive underpins the rational development of resources, and associated research provides warnings of potential over-development.

The need for a groundwater level archive derives from a number of statutory obligations placed upon government. The Secretary of State for the Environment has a duty to publish information relating to the demand for water and the availability of water resources. Without the archive, government would have to assemble information afresh to meet each new requirement for data.

The BGS has a statutory right to water data for boreholes drilled over 50 feet (15 m) (Water Resources Act 1991; a revision of the Water Act of 1945).

## History of the monitoring archive

Records show that some local water engineers and private individuals began keeping valuable records of groundwater levels as far back as the nineteenth century. Falls in groundwater levels in the Chalk of the London Basin were reported as early as the mid-nineteenth century (Clutterbuck 1850; Lucas 1877). The Institute of Geological Sciences (IGS) started to measure groundwater levels systematically in England and Wales during the 1950s.

The Water Resources Act of 1963 placed the assessment of water resources in England and Wales as a matter for national consideration. River Authorities were created by the Act, and one of their functions was to form a network of groundwater observation boreholes (this was required by the Water Resources Board). The plan was to use existing shafts and boreholes or to construct new boreholes. By March 1974, only 20 of the 29 River Authorities had submitted complete or partial schemes for such networks; six had no significant groundwater resources hence networks were not required in these areas. The remaining three authorities failed to forward any type of scheme. Observation boreholes in use at this time totalled 1393.

The Water Act of 1973 disbanded the existing River Authorities and the Water Resources Board. Ten new water authorities were created in England and Wales, together with the Water Data Unit (WDU) which became responsible for maintaining a national archive of groundwater levels. By 1980 the number of operational observation boreholes had risen to nearly 3000.

The Water Directorate of the Department of the Environment requested the Institute of Geological Sciences to review the groundwater

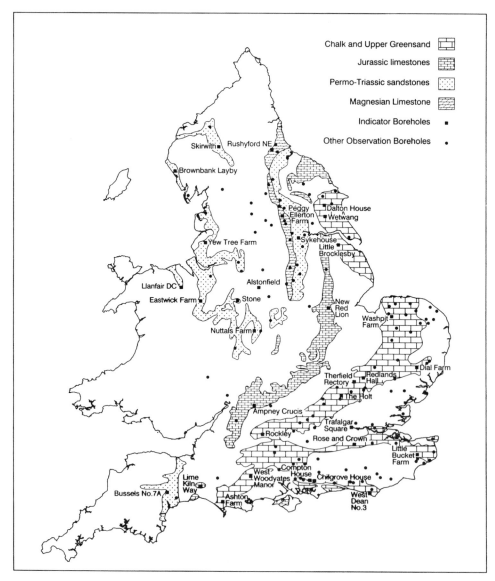

**Fig. 1.** The observation borehole network in England and Wales. The indicator boreholes are located within the principal aquifers.

level observation borehole network. The intention was to select some 200 sites from the 3000 available that might be used for periodic assessment of the national groundwater resources. Selection was based upon the hydrogeological units identified in an investigation of the groundwater resources of the United Kingdom that had been prepared for the European Economic Community (Monkhouse & Richards 1982). The units are defined as a suitable volume of aquifer within which the flow of groundwater can be considered a hydrological entity. In England and Wales, 184 units were identified and it was recommended that one site be selected for each unit. The number of sites finally selected was 175 (Monkhouse & Murti 1981). To ensure the continuing utility of the overall network it was recognized that occasional reviews would be of benefit.

The responsibility for maintaining and updating this unique archive of selected boreholes reverted to the BGS in 1980. Some sites have had to be taken out of service over the years and other new ones have been brought into use. The number of observation boreholes archived for 1993 was 161, of which 50% were in the Chalk and 24% in the Permo-Triassic sandstones. The remaining 26% of the boreholes are in the locally important and minor aquifers (Fig. 1). Three sites are located in Northern Ireland and one in Scotland.

The BGS inherited the original 3000 boreholes from the Water Data Unit in 1980, and they now form the 'Historic Groundwater Level' archive. The current 161 operational boreholes form the 'Current Groundwater Level' archive; part of the Observation Boreholes Network. The Historic Groundwater Level archive has not been updated since 1975.

## The observation borehole network

There are basically two types of groundwater observation network: local and national. Generally a local observation borehole is intended for monitoring groundwater levels in a particular area. The purpose is to observe the effects of pumping groundwater near and around major abstraction sites. These boreholes are often in temporary use; for example, when the aquifer characteristics are to be determined from a pumping test. However, in coastal regions long-term monitoring boreholes may be used to control pumping regimes in order to avoid major seawater intrusion. For national monitoring, historical records held for an observation borehole are important. Sites with less than 10 years of records have only limited value. Within the observation borehole network there are at present a considerable number of sites which have records for some 20 years, and a few with records for over 50 years. Fortunately, there are a number of sites with records extending back over 100 years. One unique example is the site at Chilgrove House borehole in the Chalk of the South Downs (National Grid Reference SU 8356 1440). The site shows an essentially unbroken record from 1836 to the present day. This is the longest continuous record of groundwater level data known to be available worldwide.

Several features at the Chilgrove House borehole are clearly outlined by inspecting the long-term hydrograph. In the months April to September, there is little infiltration since the potential evapotranspiration is generally sufficient to take all the effective rainfall; through this period the groundwater level falls. Direct recharge from rainfall takes place during the winter months during which the groundwater level rises. This is usually taken to be from October to March. The severity of the drought from 1989 to 1992 is quiet evident but, even during this four-year drought, there was a seven-week period from January to February 1990 when there was a rise in groundwater level of some 40 m in response to heavy rainfall. Subsequently, in January 1994 artesian conditions were reached and the borehole began overflowing for the first time since 1960.

A Chilgrove House hydrograph is illustrated in Fig. 2 as a solid trace for the first seven years and the last seven years of record. The record shows monthly maximum and minimum flows together with the long-term monthly average. The maximum, minimum and mean values are calculated from the years 1836–1989.

Other sites may show changes caused by human activity, for example, at Eastwick Farm in the Triassic Sandstone of the northern Welsh Borderland (National Grid Reference SJ 3814 3831). Variations in groundwater abstraction over the past ten years show a drawdown of 1.0 m in groundwater level. If this trend continues it may well affect groundwater resources in the area.

The Therfield Rectory borehole, situated in the Chalk at Royston in Hertfordshire (National Grid Reference TL 3330 3720) shows a long lag-time between rainfall and water level rise. The lag-time is about three months. Such sites are

Site Name  **CHILGROVE HOUSE**

National Grid reference  **SU 8356 1440**

Aquifer  **CHALK** and **UPPER GREENSAND**

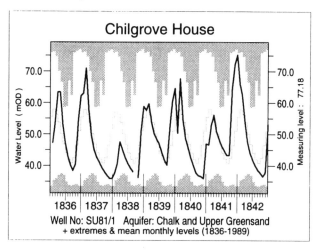

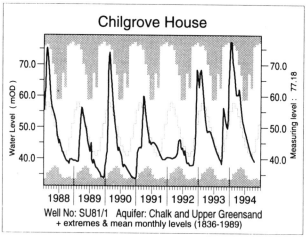

**Fig. 2**. Part of the long-term hydrograph from Chilgrove House shown as a solid trace from 1836 to 1842 and from 1988 to 1994. The maximum, minimum and mean values are calculated from 1836 to 1989. The upper and lower shaded areas show the maximum and minimum values respectively. The dotted trace shows the mean values.

valuable when considering long-term effects but it is not unusual for minor, and often interesting, responses to be blurred, and they have limited value for rapid short-term assessment. The response of boreholes in confined aquifers is similarly muted. Measurements at the site started in 1883, but groundwater levels declined beyond the bottom of the borehole on a number of occasions. Therefore the Therfield Rectory site is not wholly satisfactory for observation purposes. The site is, however, of historical value because of its long-term record. The site demonstrates some of the problems of validating very old data. Groundwater level measurements were carried out by individuals at the rectory during the early years and the instruments used, known as 'lines', were found to be unreliable. During the first 38 years the original lines had stretched by several metres. The lines were replaced in 1921, but when checked in May 1949, in contrast, the lines had shrunk by several metres.

In 1956 the borehole was equipped with reliable monitoring instruments. The ground-water data for the years up until then have been corrected proportionally for stretch and shrink-age in consultation with the National Rivers Authority. These data are not wholly reliable since the stretch and shrinkage do not occur at a uniform rate. However, the data are valuable for comparison of level changes from one year to the next and are an indication of the magnitude of those changes.

Despite the recent drought, levels in the confined Chalk aquifer below London have risen substantially over the last 20 years. Annual mean levels in the Trafalgar Square borehole (National Grid Reference TQ 2996 8051) show a 30 m rise since the mid-1960s. This is principally a consequence of reduced abstraction due to the decline of water intensive industries in the area. The decreased rate of groundwater abstraction initially stabilized the water-table which had been lowered steadily over the preceeding 150 years in response to London's water demands, and subsequently levels have risen at the rate of approximately 1 m per year.

## Data acquisition and archiving

In England and Wales the responsibility for the measurement of groundwater levels is under-taken by the water companies and the National Rivers Authority (NRA). The provision of data to the archive rests principally with the NRA. In Northern Ireland data are collected by the Department of the Environment and in Scot-land by the seven River Purification Boards.

Data are normally sought and received for the majority of sites at six-monthly intervals. Data measurements are taken either weekly or monthly, therefore the annual growth of the archive is relative to the number of water level supplied per year. The data are manually entered by BGS (Wallingford) into an ORACLE data-base with paper copies archived for security. The levels are referenced to the Ordnance Datum with a numbering system based on the National Grid.

For a number of key sites, known as the 'Indicator boreholes' (Fig. 1), groundwater levels are requested monthly. The purpose and principal function of these sites is to provide the basic information for the groundwater input to the hydrological monitoring programme for the monthly production of the 'Hydrological Sum-mary for Great Britain'.

As well as the groundwater level archive in Wallingford there are three other archives which form part of the hydrogeological database. These are the Physical Properties Archive, the

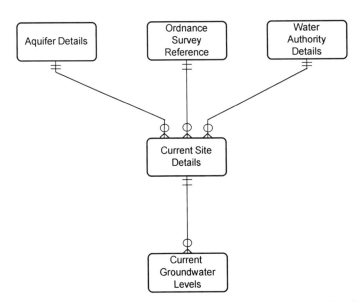

**Fig. 3.** An entity-relationship diagram for the current groundwater level archive. The historic archives are on separate but identical databases.

Hydrogeochemical Archive and the greater part of the database is the National Well Record Archive. The latter archive consists of a paper record of hydrogeological, geological, borehole and well-construction information. The approximate number of records exceed 150 000, with about 750 new sites registered annually.

Groundwater level data are stored in a system of seven tables on the ORACLE database (Fig. 3):

- current site details;
- current groundwater levels;
- historic site details;
- historic groundwater levels;
- aquifer details;
- water authority details;
- Ordnance Survey references.

The current and historic archives are kept on separate but identical databases. For each site, 20 fields of data are stored. The rest of the tables have three or four data fields relevant to the table.

The site details table stores information for each borehole. For example,the BGS reference number is a text field based on the National Grid. The station name is a text field with a width of 50 characters. The datum point and the depth of the borehole is a numeric field with a width of six digits.

Separate tables are used for groundwater levels linked to the site details table by the BGS reference number. The field for the date and time of the groundwater level measurement has a width of seven digits. The groundwater level measurement field has a width of six digits with the data stored to two decimal places in metres AOD. An accuracy field is available with a width of one digit.

Look-up tables are used to store details of Water Authorities, aquifer details and the National Grid Reference 100 km squares as a facility for country reference.

The size of the database is 4 megabytes and the data are stored on a UNIX database server at Wallingford. Access time for abstracting data is negligible. NERC authorized staff may have 'read only' access to the archive. The representative current archive has over 112 000 records, while the historic archive has over 115 000 records.

## Data dissemination

Data are disseminated through a series of publications and reports, supported by a comprehensive data retrieval service. A suite of retrieval programs has been written in order to facilitate data usage, including the production of well hydrographs in graphic form. Details of the standard retrievals are supplied on request to the public and also appear in the 'Hydrological Data UK' report series.

## Publications

### Hydrological Data UK series

Publications are produced in collaboration with the Institute of Hydrology's National Surface Water Archive Section at Wallingford.

The annual publication of a *Hydrological Data UK Yearbook* (Anon. 1993) brings together surface-water archive and groundwater archive datasets relating to riverflows, groundwater levels, aerial rainfall and water quality across the UK. The groundwater contribution includes not only hydrographs showing the fluctuations of groundwater level for the particular year of report, plus four preceding years, but also assessments of recharge received by the major aquifers.

### Hydrometric Register and Statistics

Every five years the Hydrological Data UK series includes a catalogue of riverflow gauging stations and groundwater level recording sites together with statistical summaries. The principal purpose of this publication is to provide reference material to assist archive users in the analysis and interpretation of the datasets held on the national archive. The latest five-year volume, 1986–1990, was published in 1993 (Marsh & Lees 1993).

### Hydrological Summary for Great Britain

A monthly Hydrological Summary for Great Britain report is produced. Specially selected indicator borehole sites are used to provide data for immediate assessment of the groundwater situation. Over the years 1989–1992 the monthly reports provided documentation of the drought, now shows to be the most severe this century in terms of groundwater resources.

### Occasional drought reports

The groundwater data have been used in a number of national and regional studies outlining in particular the drought situation.

Data have been used to prepare and publish the groundwater sections of the 1988–1992 drought report for the UK (Marsh *et al.* 1994) and an earlier report documenting the 1984 drought (Marsh & Lees 1985).

## Long-term hydrographs

Two long-term hydrographs have been published jointly with the National Rivers Authority. The first one was for the Chilgrove House site (Monkhouse *et al.* 1990), in the chalk of the South Downs, illustrating the groundwater level fluctuations from 1836 to the end of 1989. The second was for the Dalton Holme site (Monkhouse *et al.* 1992), in the chalk of the Yorkshire Wolds, illustrating the groundwater level fluctuation from 1889 to 1991. A third publication is in preparation for the site at Therfield Rectory, in the chalk of Hertfordshire. Data held start from 1883. The site lies approximately midway between the two sites already published, and forms a useful contribution to the hydrograph series.

## Retrievals

There are five retrieval options available. A description of each option is given in Table 1. Options 1 to 4 give details of the borehole site, the period of record available, and maximum and minimum recorded levels in addition to the output specific to each option. Data may be retrieved for a specific borehole or for groups of boreholes by site reference numbers, by area (using National Grid References), by aquifer, by hydrometric area, by measuring authority, or by any combination of these parameters.

## The future

The British Geological Survey considers the groundwater level archive to be extremely important. The current groundwater archive has operated satisfactorily for several years, and its utility was proved during the recent period of drought. Future developments of the archive are likely to be driven by two factors.

The first factor will be technological change. The rapid changes in availability of computer software will allow the integration of the database within the British Geological Survey's data architecture. This means that as one component of an Earth Sciences Geographic Information System the data can be reached rapidly and efficiently by potentially many more users. Increasingly, water levels are digitally recorded and we can look forward to the receipt of information directly at the BGS from monitoring wells in digital format.

The second factor will be an increasing awareness of the value of long-term hydrogeological data to many classes of environmental study. The recognition of the value of the data will lead to an assessment of the groundwater monitoring network at a national level. This will ensure that the required data are gathered as efficiently as possible, and that the data when collected are archived for the benefit of future research or inquiry.

## References

ANON. 1993. *Hydrological Data UK 1992 Yearbook*. Hydrological Data UK Series, The Institute of Hydrology and the British Geological Survey.

CLUTTERBUCK, J. C. 1850. On the periodical alternations, and progressive permanent depression, of the Chalk water under London. *Proceedings of the Institute of Civil Engineers*, **9**, 151–155.

LUCAS, J. 1877. The artesian system of the Thames Basin. *Royal Society Arts*, **25**, 597–604.

MARSH, T. J. & LEES, M. L. (eds) 1985. *The 1984 Drought*. Hydrological Data UK Series, The Institute of Hydrology and the British Geological Survey.

—— & —— (eds) 1993. *Hydrometric Register and Statistics 1986–90*. Hydrological Data UK Series, The Institute of Hydrology and the British Geological Survey.

——, MONKHOUSE, R. A., ARNELL, N. W., LEES M. L. & REYNARD, N. S. 1994. *The 1988–92 Drought*. Hydrological Data UK Series, The Institute of Hydrology and the British Geological Survey.

MONKHOUSE, R. A., DOORGAKANT, P., MARSH, T. J. & NEAL, R. 1992. *Long-Term Hydrograph of Groundwater Levels in the Dalton Holme Well in the Chalk of the Yorkshire Wolds*. The British Geological Survey and the National Rivers Authority, Wallchart Series No. 2.

**Table 1.** *List of groundwater level retrieval options*

| Option | Title |
| --- | --- |
| 1 | Table of groundwater levels |
| 2 | Table of annual maximum, minimum groundwater levels |
| 3 | Table of monthly maximum, minimum and mean groundwater levels |
| 4 | Hydrographs of groundwater levels |
| 5 | Site details of the borehole |

—— & NEAL, R. 1990. *Long-Term Hydrograph of Groundwater Levels in the Chilgrove House Well in the Chalk of Southern England.* The British Geological Survey and the National Rivers Authority, Wallchart Series No. 1.

—— & MURTI, P. K. 1981. *The rationalisation of groundwater observation well networks in England and Wales.* The Institute of Geological Sciences, Report No. **WD/81/1**.

—— & RICHARDS, H. J. 1982. *Groundwater resources of the United Kingdom.* Commission of the European Communities, Th. Schaefer Druckerei GmbH, Hannover.

# Database design and data management on the Swansea–Llanelli Earth Science Mapping Project

S. POWER, M. SCOTT, G. ROBINSON AND I. STATHAM

*Ove Arup & Partners, Cambrian Buildings, Mount Stuart Square, Cardiff CF1 6QP, UK*

**Abstract:** Earth science, environmental and planning information was collected for a 200 km$^2$ area centred around Swansea and Llanelli, South Wales. The aim was to use this data to produce development advice maps for use in the planning process. To ensure efficiency of data storage and retrieval, a relational database was used, combined with a geographic information system (GIS), the GIS helping with data collection and data entry. This required the design of a logical model for data storage and manipulation, together with customized database forms for input and reporting. Information was captured in a variety of forms, ranging from reports and maps to text and existing databases, and often required different input techniques. The resulting comprehensive database was therefore stored in a structured way which has enabled effective interrogation and querying of the data to produce the information around which the final maps can be drawn. These, and the data, have already proved to be of importance and interest within the study area.

## Project brief, outline and aims

In 1992, Ove Arup & Partners, Cardiff, were appointed by the Welsh Office/Department of the Environment as Contractors for the Swansea–Llanelli Earth Science Information and Mapping Project. The project brief was:

- to present relevant earth science information in a form suitable for use in forward planning, development control and environmental management; and
- to demonstrate computer techniques for input and manipulation of data, and its presentation in forms suitable for direct use.

The project is intended to provide initial guidance on earth science issues which need to be taken into account in planning and implementation of development. It is therefore mainly aimed at being for use by planners, developers and building control officers. However, direction will be given to other users on how to locate and utilize technical information which may be useful in the planning process.

A review of some of the previous similar projects (BGS 1988, 1991; Liverpool University 1990; Howland 1992; Wallace Evans Ltd 1992; Porsvig & Chrisensen 1994), and discussions with town planning staff in local authorities and the members of the project steering committee, helped to clarify the project aims. In particular it was intended to reduce the amount of technical information necessary. The result was only those data needed for the production of the maps and of direct use to planners and developers.

The brief required the production of nine thematic maps at 1:25 000 scale and an associated database of relevant information for a 200 km$^2$ area centred around Swansea and Llanelli in South Wales (Fig. 1). The area was chosen by the DoE/Welsh Office to test and develop the techniques of applied mapping adopted elsewhere in the UK to a part of South Wales which exemplifies a wide range of terrain, problematic ground conditions, and where a programme of major development and urban regeneration is being carried out.

So that the database may be put to maximum use, the datasets, i.e. the digital maps and database, are transferable onto the computer systems of the planning/engineering departments of the local authorities concerned. This encourages their use and reduces the dependency on published maps, especially since the database could be updated as new information becomes available. A brief survey was undertaken of the computer facilities of the local authorities and the results are included in the Table 1.

From Table 1, it is apparent that the results would need to be transferred to both GIS and non-GIS. No common system existed and the variety of data transfer formats and hardware systems highlighted the need for rigid data management.

## Data capture

The project required the capturing of data on the following themes:

- geological and mining information;
- minerals planning;
- potential instability;
- industrial land use;

*From* Giles, J. R. A. (ed.) 1995, *Geological Data Management*,
Geological Society Special Publication No 97, pp. 145–155.

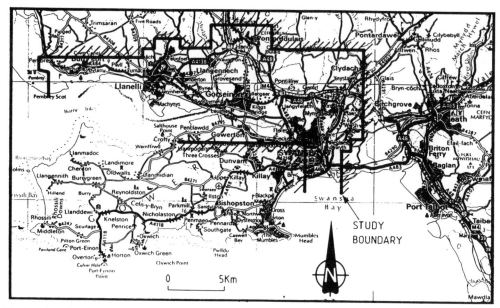

**Fig. 1.** Map of study area.

- coastal/river flooding;
- hydrology/hydrogeology;
- environmental planning;
- local and structure planning.

An initial familiarization study was undertaken to assess the volume, quantity and form of information available (Ove Arup & Partners 1992). This initial review revealed data in the formats listed in Table 2.

Potential geological database information was particularly varied, ranging from a single borehole location or stratigraphic section on a map to a site investigation report. This often contained hundreds of trial pits, boreholes, sometimes with associated soil and chemical testing and interpretative information, and historical maps. Map-based geological and land-use data were combinations of linework and closed areas.

All contained information was spatially referenced to a point or points in the study area and so grid reference was seen as the key to data storage. Hence some form of cartographic framework for data storage was considered essential.

## Hardware and software setup

Two options were considered:

- computer aided design (CAD) technology with simple links to a non-relational database, i.e. AutoCAD or Intergraphs MICRO-STATION and dBase IV; the packages were already available in-house;
- a geographic information system (GIS), e.g. Intergraph MGE/MGA linked to a relational database system such as ORACLE.

**Table 1.** *Computer facilities of local authorities*

| Authority | Databases | GIS |
| --- | --- | --- |
| West Glam CC | Yes: ORACLE | Proposed with 6 months; departmental initially ArcInfo-based |
| Swansea CC | Yes: Informix | Yes: corporate system undergoing implementation; Genamap system |
| Llanelli BC | Yes: | Yes: departmental system undergoing implementation; EDS-based |
| Dyfed CC | Yes: PC-based | Yes: departmental system operational; PC-based |
| Lliw Valley DC | Yes: Corporate | No: not proposed |

**Table 2.** *Assessment of volume, quantity and form of information*

| Data available | Examples |
| --- | --- |
| Paper maps | Old geological maps (scale 1 : 10 560) |
| | Geological maps (scale 1 : 10 000) |
| | British Coal mine entrances on plans (scale 1 : 25 000) |
| | Old Ordnance Survey county plans (scale 1 : 10 560) |
| Reports | Site investigations, containing site plans typically between 1 : 1000 to 1 : 10 000 scale |
| | Planning studies |
| | Academic studies |
| | Books |
| Databases | Aspinwall Landfill Site File (PC dBase IV system)* |
| | Liverpool University Aggregate Resources Study (PC dBase IV system)† |
| Paper printouts of databases | BGS boreholes (ASCII listing) |
| | NRA abstraction wells (full printout)‡ |
| | Welsh Office Derelict Land Survey (full printout)§ |
| | NRA/MRM sea defence survey (full printout with maps)‖ |
| CAD files | SSSIs from local authority |
| | Ordnance Survey digital contour and tidal data (scale 1 : 10 000) |

* Aspinwall (1992).
† Liverpool University (1990).
‡ National Rivers Authority.
§ Welsh Office (1992).
‖ National Rivers Authority/MRM Consulting Engineers (1991–1994).

It was felt that the latter option offered better spatial interrogation of data and, more importantly, it offered better facilities for the entry, and control of entry, of both graphic and attribute data. A 'full' GIS-based approach was also considered to be the only way to fully satisfy the second part of the brief, namely to 'demonstrate techniques for the manipulation of data'.

Since the potential client base for the data were local authorities, some of which had GIS, it was felt that we could best demonstrate the potential end uses and analysis of the data if we ourselves had a GIS package. It was felt from the outset that looking at the distribution of point items, such as mineshafts, within area features such as planning zones, would be something the end users of the data would need to do.

As an aid simply in sorting the data and retrieving information for use in map production, it is true that the CAD/dBaseIV set up or even the use of 'tagged' elements in MICRO-STATION V5 may well have been sufficient, but the degree of control in the exporting of data would not have been there.

Intergraph was simply chosen as Ove Arup & Partners have a history of CAD and civil engi-

neering design work on their hardware/software and most of the skills would be transferable to this project.

The system arrangement illustrated in Fig. 2 was adopted, based on the Intergraph 2400 series UNIX-based workstation.

The software used is listed in Table 3. This offered maximum flexibility in terms of data input formats, and range of working machines. ORACLE was mounted on the Sun workstation and accessed across the local area network (LAN).

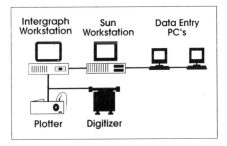

**Fig. 2.** System arrangement for project.

**Table 3.** *Summary of Software*

| Product/ Package | | Description |
|---|---|---|
| Intergraph Product | MGE | (GIS core product) |
| | MGA | (GIS spatial analysis product) |
| | Microstation | (CAD package) |
| | IRASB | (GIS package to view raster images) |
| | IPLOT | (Plotting package) |
| ORACLE | SQL* FORMS | |
| | SQL* LOADER | Database products (Relational) |
| | SQL* PLUS | |
| dBase IV | | Database product (non-relational) |
| Paradox | | Windows database product |
| Paradox | SQL* LINK | |

## Project data plan and model

A GIS package and/or a relational database requires a fairly rigid structure for the entry and storage of data. To achieve this, the following warranted consideration:

- data captured graphically are best kept in separate CAD files, and, if within one file, at least on a different 'level';
- data held on a graphical feature ('attributes') must be held in a database table, which is linked to the graphical element;
- the amount of data entered per database record, i.e. how many database tables are attached to each record?

To take account of the above, a 'Logical Data Model' was defined, in which the interrelationships of the different datasets were defined. Datasets were gathered together under broad subject categories with the category division being both a conceptual and a rigid 'Intergraph defined' structure. Each category contains a number of database tables, graphically represented by a 'feature' on a map. A feature is simply a CAD symbol or cell placed in a design file or 'map'. Each different feature is stored in a different CAD file, hence it is possible to perform both database and GIS operations on each dataset independently, and to transfer datasets as files individually.

A simplified summary for an individual data entry such as a shaft is shown in Fig. 3. In each case the same naming convention taken from the principal category (i.e. mr) is the prefix for the database table, the cell and the CAD design file.

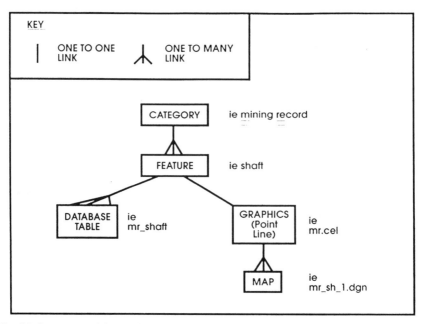

**Fig. 3.** Simplified summary of the 'Logical Data Model' for an individual data entry.

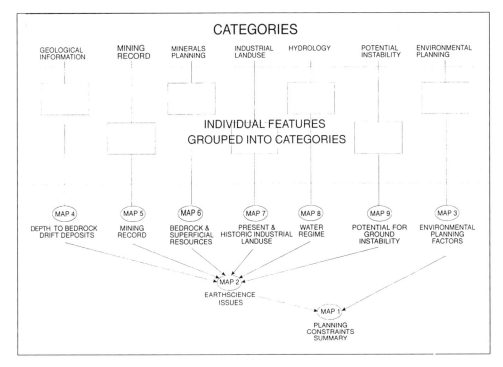

**Fig. 4.** Project structure for Swansea–Llanelli geographic information system.

In reality, with seven categories, some five to ten features per category and two or three design files for each feature, the full project data plan was considerably greater in complexity. This is illustrated in Fig. 4, and its relationship to the final maps.

## Database design and forms

The simple, single graphic feature, linked to a single database form, was the approach used extensively for the majority of the data captured. Database forms were designed and created in ORACLE SQL* FORMS. The majority were only slight modifications from the default available, with one record per page, and the ability to page up or down for the next/last record. These were basic by today's graphic user interface standards, such as 'Windows', but they were functional. A key consideration was the need to minimize the amount of typing. An example of the range and type of information entered is shown in Fig. 5.

Site investigation reports, however, required the report details and a representative sample of the boreholes and trial pits within it to be entered and displayed. A more complex form was required which allowed information from three linked tables to be displayed underneath each other on the one screen. Linkages and various triggers were set up in ORACLE to allow this to happen 'seamlessly' to the user. A 'one-to-many' relationship therefore existed, whereby a report can have none, one or several borehole records and each borehole can also contain one or many stratigraphic units. Thus it was necessary to devise linked database tables, as shown on Fig. 6.

For simplicity of data entry, the various database tables were 'hidden' behind the forms, which were set up so that prompts would assist the user to move from one form to the next. Information common to more than one database table was automatically carried over.

Items incorporated into designed forms included:

- automatic calculation of database record number/code using routine, e.g. each shaft record in the database has a unique Arup generated number;
- the allowing of only SN/SS grid references; since these were the only Ordnance Survey 100 km grid squares occupied by the study area, this minimizes input errors in this field;

● conversion of an eight-figure grid reference for a site into a twelve-figure reference required by the Microstation CAD Package to place data relative to the OS Datum, e.g. SN 2450 8560 becomes 202450, 198560; this is the x,y coordinate reference from the 0,0 grid origin within the CAD file;

● the distinction between grid references and elevations entered manually on a PC and GIS calculated grid references and elevations (occasionally different because the manually entered data come from the original SI report or borehole, which may be incorrect despite being in the correct geographic location);

SWANSEA LLANELLI STUDY——SITE INVESTIGATION DATABASE

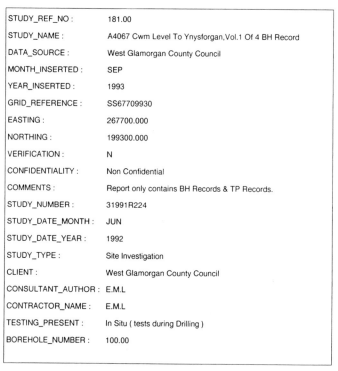

| | |
|---|---|
| STUDY_REF_NO : | 181.00 |
| STUDY_NAME : | A4067 Cwm Level To Ynysforgan,Vol.1 Of 4 BH Record |
| DATA_SOURCE : | West Glamorgan County Council |
| MONTH_INSERTED : | SEP |
| YEAR_INSERTED : | 1993 |
| GRID_REFERENCE : | SS67709930 |
| EASTING : | 267700.000 |
| NORTHING : | 199300.000 |
| VERIFICATION : | N |
| CONFIDENTIALITY : | Non Confidential |
| COMMENTS : | Report only contains BH Records & TP Records. |
| STUDY_NUMBER : | 31991R224 |
| STUDY_DATE_MONTH : | JUN |
| STUDY_DATE_YEAR : | 1992 |
| STUDY_TYPE : | Site Investigation |
| CLIENT : | West Glamorgan County Council |
| CONSULTANT_AUTHOR : | E.M.L |
| CONTRACTOR_NAME : | E.M.L |
| TESTING_PRESENT : | In Situ ( tests during Drilling ) |
| BOREHOLE_NUMBER : | 100.00 |

**Fig. 5.** Output in PARADOX of data held in ORACLE.

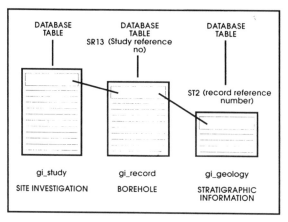

**Fig. 6.** Outline of linked tables used in ORACLE

separated linked tables containing a data dictionary of available defaults for a field in the database, e.g. a list of allowable sources, contractors, etc. The operator was forced to pick from these lists, minimizing the possibilities of erroneous data and reducing the typed data input. This was possible because a thorough review of a wide range of data entry options had been undertaken at the beginning of the project, enabling comprehensive lists of most of the fields to be produced, from geology types to types of boreholes.

## Data input techniques

A wide range of data input techniques were investigated and six key methods were used:

- manual entry into forms in ORACLE on a PC (e.g. for site investigation reports, borehole records);
- entry on screen of workstation into database via GIS menus (e.g. for stratigraphic records taken from BGS 1:10 000 scale geological maps).
- entry into dBase IV on PC, with subsequent transfer of data to ORACLE (e.g. for Welsh Office Contaminated Land Database printouts);
- scanning maps and digitizing from them, both on digitizing tables and on the workstation screen (e.g. for hydrological information from 1:10 000 OS maps and land use information from 1:10 560 historic maps);
- scanning of text into ASCII format and reading the file into the ORACLE database on the Sun workstation using SQL* LOADER (an example was the Borehole database printouts; these were not made available to us on disc at the time of the study so this was the preferred method to save typing them all in);
- importing CAD files into either Intergraph or Sun workstation in .DXF format and attaching database information (e.g. for SSSI information from a local authority).

Of the six methods, the best, giving the least problems and most control over entry, was the entry directly on screen into the GIS software of both graphics and attribute data since much could be defined in advance for the operator. All of the other methods required considerably more quality assurance and a higher level of operator/system skill.

The information imported was worked upon to provide an intelligent GIS feature with a full set of database attributes. Four key functions within the GIS software packages enabled full datasets of information to be produced relatively easily and with reduced data input. These are, by necessity, rather specific to Intergraphs MGE Software, but the general techniques are believed to be available in some form in all of the major GIS software packages. These were:

- a function that takes the grid references of a record stored in the ORACLE database table and places a point in the correct cartographic location on the screen, and then links the stored attributes to it;
- a function that will automatically calculate the area and grid reference of any features placed into a CAD file and put the subsequent calculated values into the database records;
- a function that, when run on a CAD design file, will convert points and lines into 'features' and attach a blank database record accordingly;
- a software package, IRASB, that allows the user to view scanned maps, move them and rotate them to the correct cartographic position, thus enabling much information to be digitized directly on the screen, particularly when the original data were of mixed scale or had a different cartographic projection.

Without these, a considerable amount of time would have been wasted by data entry operators on routine tasks. Instead, the amassing of 50+ attribute datasets in a structured manner has been possible by just one or two operators in six to eight months, with limited system guidance. Some datasets contain up to 1500 records.

## Quality assurance

In order to ensure that the datasets are in a form in which they can both be accessed by users and transferred to a third-party GIS or database system, rigorous quality assurance is needed.

Quality assurance procedures were adopted from the outset and some examples are discussed below.

### Data entry into ORACLE forms

Extra columns were added to the database tables so that the status of a particular record could be recorded and checked; in particular 'verification' and 'confidentiality'.

The data given in a report (or taken from a map) may contain errors at source or operator error. Thus it was necessary in the case of both grid reference and elevation to have two columns for these data attributes, the second set termed 'GIS calculated' so that any discrepancies became apparent.

A column was finally added to the database table called 'GIS Management' which enabled a character, i.e. Y/N, to be inserted into the database record once a GIS process (such as updating all grid references) had been run. Thus it was possible to search using SQL for all records that contain this particular character (i.e. 'Y') and hence 'pull out' all records affected or updated during a particular process. The contents of the GIS Management column do not therefore affect the database structure and can be removed at a later date.

### Data entry of graphical datasets

The steps undertaken from placing lines in a CAD design file to obtaining a full quality assured GIS dataset are many and form a distinct process or 'workflow'. A diagram/table was produced in which each operation in the workflow could be 'ticked off' against each of the 50 or so graphic files, (Fig. 7). Note that not all graphics files have attributes attached.

Key stages in the work flow include:

- removing any graphic elements that are not needed, i.e. overshoots and free end points, and small 'slivers' of areas caused by duplicate linework; making sure the linework that makes up any areas is closed is vital for colour filling of shapes later on;

- placing values and text into the database columns (attributes) using automatic routines; for example, in the case of 80 or so site investigation reports entered from the same source, the 'source' column could be updated to 'Llanelli Borough Council' for all 80 records in one short process;

- the printouts of the digitized information are compared manually with the original data and checked for errors; for example, by using a light table when digitizing shafts/adits from scanned maps on the screen.

- unique attributes (such as date of workings for a shaft) need to be entered as information becomes available and since this is not always in a structured manner, checking is required for progress and consistency at

FEATURE CATEGORY    Hydrology

| FEATURE NAME | DESIGN FILE | DIGITISE DATA | CLEAN LINEWORK | ADD CENTROIDS | LOAD AREAS | UPDATE/ENTER ATTRIBUTES |
|---|---|---|---|---|---|---|
| Springs | hy_sp_1 | ✓ | ✓ | N/A | N/A | N/A |
| Water wells | hy_ww_1 | ✓ | ✓ | ✓ | N/A | ✓ |
| Rivers | hy_riv_1 | ✓ | ✓ | N/A | N/A | N/A |
| Licenced Abstractions | hy_nra_1 | ✓ | ✓ | ✓ | N/A | ✓ |
| Reservoirs | hy_res_1 | ✓ | ✓ | ✓ | ✓ | ✓ |
|  |  |  |  |  |  |  |

**Fig. 7.** Example of typical quality assurance workflow for graphical datasets.

regular intervals to ensure that no records are only partially filled;

- graphic points are often by necessity placed in the centre of any area features such as a 'land use parcel' or 'geological unit'. Attribute data are attached to these points. Occasionally duplicates of these occur by virtue of software bugs and these have to be removed, otherwise they will cause problems later with queries.

### Data entry of non-graphical datasets

The entry of any information remotely on PCs lacks the benefit of the cartographic GIS facility, thus it is possible that as a result of errors occurring in entering grid references, records are misplotted.

Plotting non-graphical datasets, such as the Liverpool University dBase IV study, graphically, using the GIS proved to be very useful, for example in removing the occasional obviously incorrect database record (i.e. boreholes/landfill sites) from the sea.

### Graphic and non graphic record consistency

Data entered into database tables may not have an associated attached graphical feature and conversely graphical elements can exist without data attached. This is often because the necessary processes were not always run on the full dataset or in the correct order. This can easily be checked and rectified during routine QA procedures.

## Data transfer

The resultant quality assured data can be exported as a series of comma delineated ASCII files (containing attribute data) which, together with CAD files in .DXF format (containing graphical features), can be transferred in stages to a third-party GIS or CAD system.

The attribute data, as an ASCII file, are taken into either a pc-based database package, such as Microsoft Access or DBase IV, then into ORACLE/Informix. The graphical data can be taken into packages such as AutoCAD using standard 'DXFin' functions.

The linking of the two back together is the key to getting maximum functionality from the data. One local authority has achieved this using a 'hookup' function in their GIS software. If this full re-linking of the data is not possible by the local authorities it is hoped that the CAD files will

at least be used as a reference in their GIS and that the attribute data will reside in a PC database to be queried in the standard, non-graphical way.

## Map design

The collection and organization of the project datasets, while initially governed by the Project Structure (Fig. 4) was further streamlined and organized by closely following an agreed set of map keys.

The content of the keys, their style and level of detail, was agreed at a series of consultative meetings held during the project's duration. They clearly set out very early on what needed to be captured graphically and in what level of detail. An example of a map key is shown in Fig. 8.

Key elements are the guidelines/legislation/action and the consideration section. All are designed to take the earth science information into the planning departments and ensure that it is understood, its relevance to the recent legal legislation is apparent, and that the implications of development without at least investigating the factors in more detail is given. The aim is for the maps to act as the first point of reference, and in many cases to give initial data that can be taken up with the appropriate statutory consultees (NRA, British Coal, etc.).

Factual data were mapped at a scale of 1:10 000, entered into the computer at this scale and, as mentioned, brought into the final maps at 1:25 000 scale. There is an element, therefore, of improving accuracy in the capturing of the data, by doing the base work at one scale higher than the output.

Information from the database was used largely to produce labelled points, so that the distribution of key database attributes, such as 'type of licenced abstractions' or 'superficial deposit type in top layer of borehole' could be displayed and plotted. These were then used to act as a base for the hand interpretation of the earth science factors, such as shallow mining hazard zones, areas of groundwater abstraction, areas of flooding and the geological based map areas.

Most of the geological linework, (seams, geological unit boundaries, etc.) was only digitized after the interpreted amended information was checked. This reduced the need to edit complex graphical shapes on screen.

Although initially it was intended to capture most, if not all, of the linework present on a traditional geological map, detailed discussion with the end users of the maps revealed that much of the geological information, such as

## MAP 2 : EARTH SCIENCE FACTORS

### LEGEND

| SYMBOL | DESCRIPTION | GUIDELINES/ LEGISLATION | ACTION | REPORT SECTION | CONSIDERATIONS |
|---|---|---|---|---|---|
| | Potential for ground instability due to shallow mining / Potential for ground instability due to land slipping | PPG 14 - Development on Unstable Land, 1990. PPG 20 - Coastal Planning, 1992. | Consult Building Control. Consult British Coal. DoE Natural Cavities Study. Stability Report required | Map 9 and Chapter 3 of Report | No technical restrictions. Subsidence / Subsidence / Possibility of development restrictions depending on scheme |
| | Near surface compressible strata | PPG 14 Development of Unstable Land, 1990 | Site Investigation Required | Map 4 Chapter 4 | Ground Lowering / No technical restriction |
| | Area of potential flooding ( coastal or river ) | Circular 17/82 - Development in Flood risk Areas - Liaison between Planning Authorities and Water Authorities. Consultation Paper - Development in Flood Risk Areas, 1992. PPG 20 coastal planning 1992 | Consult NRA | Map 8 Chapter 5 | Possible technical restrictions construct adequate level defences |
| | Present and Industrial land use: areas where existing/previous land use need to be considered / Waste disposal sites | Circular 20/92 - Planning Controls for hazardous substances. Consultation Paper - Planning and Pollution Control, 1992. Draft for Consultation, Waste Management Paper No. 2/3 - 'The Preparation of Waste Disposal (Management Plans) | Consult Planning, Environmental Health, Building Control and Borough Surveyors Departments. Desk Study required as a minimum, possible chemical testing. Investigation Required Consult Environmental Health Department of the appropriate Local Authority | Map 7 Chapter 6 | No technical restrictions / Potential Contamination / Leachate polluting watercourse |
| | Made Ground | PP14 - Development on Unstable Land, 1990 | Consult Planning, Environmental Health and Borour Surveyors Department | Map 4 Chapter 4 | No technical restrictions |
| | Minerals : Areas where mineral extractions should be considered | Policy on the protection of ground water NRA (1982) Water Resources Act Environmental Protection Act | Consult NRA | Map 6 Chapter 7 | No technical restrictions. Surface/Groundwater Protection |

Fig. 8. Examples of a map key.

unconformities and stratigraphic boundaries, was surplus to requirement.

The decision was taken to capture only those areas of drift geology that warranted consideration in the planning process, and none of the solid geology. Only the location of coal seams was deemed important enough to remain.

Of the remaining data, none was attributed. The amount of value-added information that could be added to an area of 'low compressibility clay' for example, that would be of meaning/relevant to the planners was negligible. Instead, the decision was made to supplement the maps of drift/coal seams with detailed annotation and labelling.

## Conclusions

Adopting and using a good CAD product like Microstation has meant that the maps have been produced relatively easily and quickly, and are of good quality. In addition, amendments and producing 'derivative' compilation maps taking data straight from other maps has been easier. A database was definitely needed for the storing of much data identified during the study, and the ability during map production to identify and recall data on a mapped item was vital.

'True' GIS analysis, however, such as consideration of overlapping areas, zoning and proximity analysis, was not undertaken to any large extent. Neither was substantial direct interpretation or map production from the database, or reporting using the GIS in-built packages. The two principal reasons are:

- the analysis of the data and its detailed comparison against planning data is the task of the end user of the data;
- geological mapping is generally too complex to allow automatic and direct interpretation from data held in a database, unless an extraordinary amount of data is entered and coded at a very close spatial placing; even then, much effort would be expended in trying to model using the computer data.

The maps have been well received within the study area and it is anticipated that particularly within the two urban authorities of Swansea and Llanelli they will become a standard tool in the preparation of the local plan documentation in the next few years.

Interest has been shown by development control planning officers in using the maps at the 'screening' stage in the assessment of planning applications in the search for potential constraints. At a time of increasing loss of local knowledge, local government reorganization and reorganization in the near future of some of statutory consultees, the results of the project will hopefully gain importance with time.

Maintenance and updating of the database and maps is not part of the contract but it is anticipated that the interested parties in the local authorities will take up this task. It is the monitoring and feedback on their use that will determine the future direction in which work of this sort is undertaken.

The authors wish to thank W. Hoare, C. Toms, H. Mordecai, R. Cottrell and numerous others who have contributed to both the research and the project work. This paper has been prepared in the course of research supported by the Welsh Office/Department of Environment, Contract No. PECD 7/1/427.

## References

ASPINWALL 1992. *Site File: Landfill Sites Database*. Aspinwall Technical Report and Database (held in Welsh Office).

BGS 1988. Deeside (North Wales). *Thematic Geological Mapping*. BGS Technical Report **WAS/88/2**.

BGS 1991. Applied Geological Mapping in the Wrexham Area, Geology and Land Use Planning. BGS Technical Report **WAS/91/4**.

HOWLAND, A. F. 1992. The use of computers in the Engineering geology of Londons Docklands. *Quarterly Journal of Engineering Geology*, **25**, 257-267.

LIVERPOOL UNIVERSITY 1990. *An Appraisal of the Land Based Sand and Gravel Resources of South Wales*. Draft report to Welsh Office/DoE.

NATIONAL RIVERS AUTHORITY. *Database of Licensed Well Abstractions*. Technical Report NRA SE, Wales Regional Office.

NATIONAL RIVERS AUTHORITY/MRM CONSULTING ENGINEERS 1991–1994. *Survey of Sea Defences in Wales*. Technical Report and Database, NRA SE Wales Regional Office.

OVE ARUP & PARTNERS 1992. *Familiarisation Study—Swansea–Llanelli Earth Science and Mapping Project*. Report No. **92/2143**.

——1993. *Information Coding, Storage and Output—Swansea–Llanelli Earth Science Information and Mapping Project*. Discussion Paper, Report No. **93/2218**.

——1994. *Islwyn Shallow Mining*. Draft Report No. **94/2426**.

PORSVIG, M. & CHRISENSEN, F. M. 1994. The Geomodel: the geological and geotechnical information system used in the Great Belt project. *Proc. Instn. Civ. Engrs Geotech. Engng*, **107**, 193–206.

WALLACE EVANS LTD 1992. *Presentation of Earth Science Information for Planning Development and Conservation, The Severn Levels*. Draft reports.

WELSH OFFICE 1992. *Welsh Office Survey of Contaminated Land in Wales, and Database Report*. Welsh Office/Liverpool University, EAU.

# A hydrogeological database for Honduras

ANDREW A. McKENZIE

*British Geological Survey, Maclean Building, Crowmarsh Gifford*
*Wallingford, Oxon, OX10 8BB, UK*

**Abstract:** An information system for the management of groundwater data was developed for the government of Honduras. The database design concentrated on close attention to the requirements and skills of the users. The database was implemented using commercially available database languages and programming tools. This allowed a bilingual system closely matched to local requirements to be developed at low cost.

The UK/Honduras Groundwater Investigation and Development project is a programme of technical co-operation funded by the Overseas Development Administration. Some 30% of public water supply in Honduras is derived from groundwater. In medium sized urban areas, the chief focus of the project, this may rise to 50%. These resources had been developed piecemeal over several years, generally without adequate study or documentation.

Since the mid-1980s, hydrogeologists seconded from the British Geological Survey have attempted to redress this shortcoming. They have worked with civil engineers from the Honduran State Water Company, have studied the nation's groundwater resources and helped to set up a groundwater investigation unit.

One major project task, begun in 1989, was to set up a digital database of hydrogeological information that could be used to manage the extensive but disorganized paper records. This also allows rapid appraisal of the groundwater resources of an area without labourious searching of paper archives. This paper documents that database and highlights the issues involved in its design and structure.

The use of trade names in this paper does not imply endorsement of the named products by either the British Geological Survey or the Natural Environment Research Council.

## The Dataset

The work of the groundwater investigation unit involves several different activities. The unit is primarily responsible for advising other parts of the State Water company, other government departments, aid projects and institutions on the probability of obtaining groundwater supplies, and where appropriate, siting new wells and supervising their drilling and testing. It advises on the management, maintenance and operation of existing groundwater-based water-supply systems, and introduces engineers to appropriate hydrogeological techniques. It also plays a major role as a clearing house for hydrogeological information at a national level.

To support this work the unit requires rapid access to information, and to be able to distribute this information in a form that will be easily understood by planners and engineers. The unit manages an archive of groundwater information and publishes hydrogeological maps. All these activities are underpinned by a digital database of information on drilled wells.

The data requirements of the unit were examined during the early months of the project. This initial study was used to decide the essential information that required incorporation. The careful attention paid at an early stage to user requirements was a key element in the project's success.

The final user requirement was for a database to store information on:

- well location;
- well construction;
- pumping equipment;
- well condition;
- pump tests;
- water levels;
- field chemical analysis of water samples;
- laboratory chemical analysis of water samples;
- reports concerning wells.

## Choice of software

After initial experiments with databases designed and sold specifically for hydrogeological information, the decision was taken to set up a customized database using general-purpose commercial relational database software. This approach has been followed by several authors in different countries and hydrogeological environments, (Dikshit & Ramaseshan 1989; Menggui Jin *et al.* 1993).

*From* Giles, J. R. A. (ed.) 1995, *Geological Data Management*,
Geological Society Special Publication No 97, pp. 157–162.

This decision was influenced by many factors, but three especially.

- The existing bespoke databases for hydrogeological information did not store appropriate descriptors needed by the water authority and appropriate to Honduran conditions. Rigid table structures meant that irrelevant data fields were presented to users, and some vital local data were impossible to store.
- Verbose user interfaces and the use of complex terms made the existing software hard to use for non-English-speaking engineers with little prior computer experience. There was a need for a bilingual system.
- The inability to import existing data from previous projects was considered a serious limitation.

The software chosen was Software Publishing's Superbase for Windows™ (Anon1992), a relational database that works within Microsoft Windows 3.1™ on IBM PC compatible computers. Some supporting software was written in Microsoft Visual Basic™.

The factors influencing this choice of software were:

- *availability of a graphic user interface*: the consistent user interface provided by Windows considerably simplified the training of Honduran users in the operation of suites of software and provided a programming environment that made writing modular programs that exchange data simple;
- *ability to import existing data*: before development of the project's own database began, a not insubstantial amount of data had already been entered on computer using a variety of software packages, ranging from spreadsheets to specialist hydrogeological programs.
- *Maintenance and support*: by using a well-documented commercial package it was possible to localize both development and support, ensuring rapid responses to user requests for modifications and improvements.

The use of Microsoft Windows demands powerful computers, but the database adopted operates acceptably on 80286 computers, given adequate memory.

## Database structure

The database was designed from the outset to store a basic set of information required by the project engineers to make rapid assessments of groundwater resources and borehole productivity region by region. From the outset it was envisaged that the database would act as controller to specialized programs that can store and display detailed information. Comprehensive records are still kept as parallel paper archives.

The final design of the database was arrived at by a process of refinement over three years, with little reference to published work outside the manuals for the software and the popular computer press. With hindsight, several elements of the design could have been improved.

Although the database now only records information about drilled wells, it was decided that it should be structured so that in the future it could be expanded to include other data needed by the project. This might take the form of information on surface water, springs, handdug wells, meteorological stations or engineering works.

The database was set up as several linked tables. All stations are entered into a location table, which provides the core of the database. A unique identifier is assigned to each station. The unique identifier is made up of an initial string identifying the administrative region, then an identifier for the type of data, and finally an accession number. This numbering scheme has the advantage of allowing rapid geographic searches, but makes updating complicated if a station's administrative area requires revision due to administrative changes or receipt of accurate coordinates.

For details other than location, several subsidiary tables are used. Figure 1 shows the principal tables and their relationships.

For drilled wells, the details of construction and usage are recorded in a single table. This table records construction details such as depth, casing type, casing location, screen depth, screen location, date of drilling and ownership. This information is complemented by a review of the last known status of the well and a freeform memo.

All wells have an entry in the location and construction files but other data, such as water quality or pump test results, are stored in their respective files only if data are available. Generally there is a 'one-to-many' relationship between wells and these files of hydrogeological detail. This presented some problems in reporting and viewing summaries of the data in the first version of the software, so provision was made, by means of a summary file, for the user to identify the 'best' analysis or test, which can be used in views and reports. Later versions of the software have improved reporting capabilities,

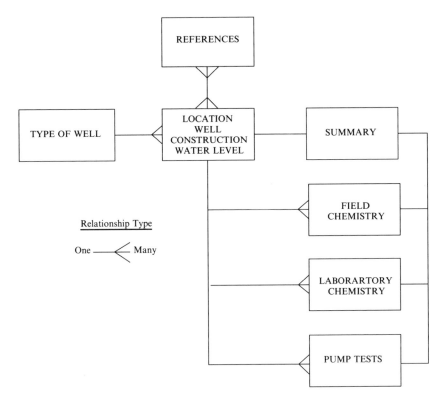

**Fig. 1.** Principal tables used by the database system and their relationships.

but it is still preferred to identify a 'best' analysis, rather than average data of varying reliability.

The most complex relationship encountered was that of the location of wells and the bibliographic database of published and internal reports. A separate table was used to maintain the relationship between these files, but limitations in the display of transactions in early versions of the database software made the handling of these data cumbersome.

Neither complex time series data nor detailed lithology are stored in the database. It was judged that handling these data would impose unacceptable demands on searches and that it would not be possible to provide suitable (graphical) output for this data from within the database. Instead the database includes an index to the location of the detailed data.

Where these data are in digital form on the same computer as the database, the software can initiate a suitable data display or analysis program and transfer the user focus to that program. Figure 2 summarizes the relationships between the database and external programs. This initiation facility considerably simplified the database programmer's task, and allowed the use of a variety of specialized programs, both those available commercially and those written specifically for the project. One important benefit from this approach is that updates to the external programs do not require the rewriting of database modules. Communication between program modules was kept as simple as practicable. Generally, on user selection of an external program, data is transferred by ASCII files and commands by keystroke emulation. DDE (Windows Dynamic Data Exchange, allowing two Windows$^{TM}$ programs to exchange data) or OLE (Object Linking and Embedding, allowing a Windows$^{TM}$ program to include a link to a second program within its data files) links between programs would provide improved performance but at the expense of greater programming complexity.

## The user interface

Considerable emphasis was placed on the user interface, as in many respects it was limitations

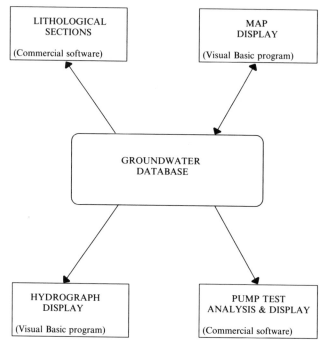

**Fig. 2.** Relationships between the database and external programs used for manipulation and graphical display of detailed data.

in the available program's interfaces that prompted the development of the application.

The database was designed to be used by engineers with limited general experience of computing. While all the Honduran project staff preferred to work with Spanish language programs, it was recognized that the UK staff would prefer an English interface.

On starting the system the user is presented with a choice between the English and Spanish versions of the program. The underlying data structure and database files are identical, but the user forms and the dialogue boxes and prompts are altered for each language. Language-specific information was stored in ASCII files read by the program on startup to allow easy modification or translation to other languages.

The principal form presents the user with a summary view of the data (Fig. 3). Pull-down menus allow the user to select activities such as printing, data entry, different views, geographical searches or administrative functions. The most common functions can be accessed by buttons placed directly on the form, and the form can be browsed using buttons that provide a 'tape recorder' metaphor placed at the bottom

of the screen. Screens are consistent in design and use colour conventions to distinguish between and highlight editable data, lookup data and calculated data.

Initial data entry for a well is through a form that combines location and construction data. Subsequent editing is through forms specific to the underlying database files. Summary forms allow the user to select the 'master' analysis or test when more than one is available.

Geographic searches can be based on selection of administrative area, or by coordinates. Radial searches around a point are supported, but they proved to be slow, whereas searches within a square centred on a point are faster. To provide the user with some sense of the location of wells a simple geographical display has been developed. Selected areas were scanned to provide bit map images, and the map coordinates of the image corners were recorded. When the user selects an image, a search is made for wells that lie within the specified coordinates, and they are then superimposed on the map within a window. The user can select a well from the window and return to the database. These features are provided by an independent program written in Visual Basic$^{TM}$.

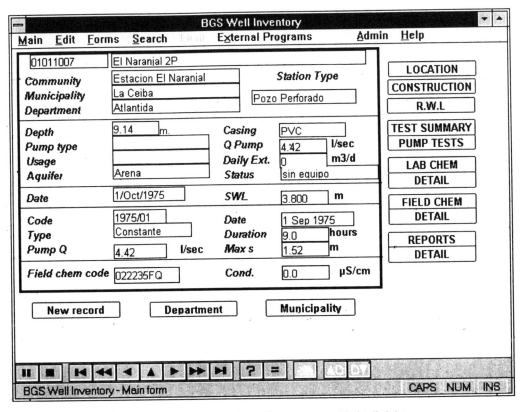

**Fig. 3.** Data browsing screen used to select records or choose screens with detailed data.

A set of standard reports can be easily selected; for more complex searches the user needs to understand the query dialogue language of Superbase™.

## Future enhancements

The software is continually upgraded to satisfy changing user requirements, and to take advantage of enhancements to the underlying commercial database software.

The recent release of the public-domain United Nations Groundwater for Windows software (Karanjac) has opened the prospect of using the graphical presentation facilities of this software to address the requirements for a graphical display of lithology and well construction. To this end, routines will be written to allow easy transfer of data between the Honduran inventory and the UN software. The Honduran inventory provides capabilities, specifically sophisticated user query writing,

customization of menus and the ability to handle multiple samples and tests for a single well, which are not yet available in the UN software.

## Conclusions

The database design is demonstrably successful because it is used on a daily basis by the hydrogeological unit of the Honduran State Water Company. It was tailored very specifically to the unit's requirements, but with a structure that should allow future enhancement or modification of the database for different requirements and use in other countries.

The use of a commercial database for this purpose is not novel, but the following lessons learnt can be applied to other similar projects.

• Database structure should be adaptable to the specific needs of the users; the types of data that will be important in one country may be considered superfluous in another.

- Analysis of the level of computer skills and language understanding of the users may be just as important as good underlying database design.
- The user interface should be designed to be as simple as practical for average users, while allowing access to more sophisticated functions where required.
- The availability of environments like Microsoft Windows™, running several processes simultaneously, allows the developer to use a relational database where appropriate, and link to other software accessing different files where this provides performance advantages. This also means that the 'suite' of software used by the hydrogeologist can be upgraded without abandoning previous investment.
- A commercial relational database package can be simpler to operate than one written

specifically for a geological market, and provides the database administrator with powerful tools for data import and export.

## References

ANON. 1992. *Developing Applications in Superbase*. Software Publishing Corporation, Santa Clara, USA.

DIKSHIT, A. K & RAMASESHAN, S. 1989. An interactive groundwater database system. *In*: *Proceedings of an International Workshop on Appropriate Methodologies for development and management of groundwater resources in developing countries*, Vol. 1. National Geophysical Research Institute, Hyderabad, India.

MENGGUI JIN, BEIYI JIN & BINHAI ZHANG 1993. Database management system for a ground-water regime. *Ground Water*, **31**, 4.

# CD-ROM and its application to the petroleum industry

STEPHEN G. ALLEN

*Pinwydden, Llanrwst Rd, Bryn-y-Maen, Colwyn Bay, Clwyd LL28 5EN, UK*

**Abstract:** Over the last few years CD-ROM has emerged in many industries and professions as an essential means of distributing and accessing information. The petroleum industry, however, has been slow to realize the potential of this medium with, until recently, only a few isolated applications available. A detailed description of the CD-ROM medium and its manufacture is provided and its relevance to the oil industry discussed. Some examples of early and current applications are presented and discussed.

The petroleum industry has often been at the leading edge of technology, specifically in the development of sophisticated workstations and software to manipulate and process information. However, despite (or perhaps because of) the vast quantities of data, current methods of storing and distributing data are still relatively antiquated with off-site storage and bulky distribution of vulnerable paper and magnetic media. The current drive towards standards led by the Petrotechnical Open Software Corporation (POSC), the American Petroleum Institute (API), the American Association of Petroleum Geologists (AAPG) and others will undoubtedly result in a further increase in the demand for and use of digital information. The explosion in data exchange requires the adoption of a cost-effective and durable medium on which data can be stored for long periods yet be accessible to many different users on different systems. The following discussion will, it is hoped, provide convincing evidence that CD-ROM represents such a medium.

Increasingly CD-ROMs are appearing as both an inexpensive and innovative distribution and publication medium and as a standard, durable and efficient means of storing and archiving data for the future. Those aspects that make CD-ROM such an exciting medium for the future are referred to in the text with examples of how these are being applied. Examples of some current applications are briefly described and used to demonstrate the versatility and opportunities which CD-ROM has to offer the petroleum industry both today and in the future. A detailed description of the history and development of the CD-ROM and its method of production is provided in Appendix 1.

## What is CD-ROM?

Many people are no doubt already familiar with compact discs (CDs) and CD-ROMs, particularly in the domestic audio industry. The author has personally used many CD-ROM applications and participated in the preparation of several others. However, it was only through research for this paper that the author really began to understand about CD-ROM, its origins and manufacture.

## Why use CD-ROM?

CD-ROMs could only be produced by specialist companies until recently. All the data and software to be placed on the disc had to be prepared and a single image of the entire disc created. This image was then burnt into the master disc from which hundreds or thousands of copies are replicated. The process is costly and requires forethought to ensure that all the data are prepared and that no mistakes are present. As with publishing a book, corrections cannot be made once the disc has been made without completely re-publishing.

In view of the apparent cost and inconvenience of the production procedure it is not surprising that many sceptics have asked why use CD-ROMs at all. It is, however, the very nature of CD-ROMs and their production process that give this medium such unique properties, creating opportunities for data archiving and distribution previously unattainable.

## Open systems

Optical discs date back to the 1960s when laser technology was first demonstrated; in the 1980s WORM (Write-Once Read-Many) discs were developed primarily as archiving or back-up systems. As the acronym suggests, these data could not be erased from the disc and were therefore ideal for archiving. However, the lack of standardization and variety of formats and media sizes, undermined the value of such discs, and many of the early versions are now no longer supported.

*From* Giles, J. R. A. (ed.) 1995, *Geological Data Management*,
Geological Society Special Publication No 97, pp. 163–180.

Later developments produced the REO (Removable Erasable Optical) or MO (Magneto-Optical) disc. These differed from WORM discs by using a combination of optical and magnetic technology enabling data to be repeatedly written and erased from the disc. The REO technology has ultimately spawned a whole range of products aimed at providing high-capacity removable discs. These discs range in capacity from 20 megabytes to hundreds of megabytes or even gigabytes. Lack of standards has, nevertheless, prevented these discs being utilized for distribution or exchange of information between companies with differing hardware platforms and drives.

CD-ROMs have their roots in the audio industry where, in the early 1980s, laser technology in the form of CDs was employed to replace traditional records and cassette tapes. CDs enabled high-quality digital music to be distributed on a durable media virtually immune from damage in what is regarded as one of the harshest environments, the domestic consumer. In 1984, Phillips Inc. promoted the concept of using CDs to store and distribute digital data for use with computers; the CD-ROM as we know it was born. Eager to gain rapid industry acceptance of the media, an early agreement regarding the precise method of encoding and decoding data from the CD-ROM was essential. In May 1985 a group of interested parties including software, hardware and media producers met to define the CD-ROM and how data should be stored on it. The result became know as the High Sierra Format, named after the High Sierra Hotel where the meeting was held, and the participating companies agreed to produce CD-ROMs, readers and software conforming to this convention. Not all companies were prepared immediately to adopt this format, however, and several alternatives co-existed, including Apple and Hitachi, through the mid to late 1980s. In 1988 the High Sierra Format finally became accepted as the standard and, with some slight modifications, was ratified by the International Standards Organisation in 1989 and published as the ISO 9660 format.

The ISO 9660 format determined the physical dimensions and characteristics of the CD-ROM, making all CD-ROM drives compatible. Of equal importance, the format established a method of encoding data that was independent of the hardware and operating system used to create it. Developers could be confident that the CD-ROMs they produced could be read in any CD-ROM drive attached to any type of computer and operating system supporting the ISO 9660 format definition.

Since the early days many hundreds of CD-ROM titles have been produced in many languages and covering a wide spectrum of subjects and applications. CD-ROMs themselves are readable across computer platforms; however, executable programs written for specific operating systems, for example MS-DOS®, and published on the CD-ROM, will not be operable on other operating systems such as UNIX™. Care and forethought are therefore required before publication, to determine the potential market for the CD-ROM and the targeted operating systems to be supported. If only data files in standard formats are placed on the CD-ROM, the retrieval and display software can be distributed on other media such as floppy discs appropriate for the platform required.

This internationally agreed standard format, ISO 9660, enables data vendors and distributors to prepare one set of media that is readable by a wide range of CD-ROM drives, host computers and operating systems. This also ensures that, not only will the CD-ROM physically survive well into the 21st century, but also the drives and interfaces capable of reading CD-ROM discs should also continue to be available. Data archived today will still be readable tomorrow despite advances in computers and operating systems.

*Durability*

One of the most significant features of CD-ROM is directly attributable to its manufacturing process. Data are physically and permanently stamped into tough polycarbonate making the CD-ROM virtually immune to normal changes in heat, light, humidity and shock. Since no magnetic material is used the CD-ROM cannot be corrupted or damaged by magnetic fields and other forms of radiation. The reading head is, in relative terms, a considerable distance from the surface making 'head-crashes' virtually impossible. Constant re-focusing of the laser beam on the reflective surface enables minor surface scratches and distortions of the CD-ROM to be effectively ignored. The result is a very impressive life for the media of over 30 years. This durability makes CD-ROM ideal for long-term storage or archival of data. Since there is no requirement for special storage conditions, and given the remarkable compactness of the medium, large quantities of information can be stored in a very cost-effective manner.

CD-ROMs are inherently non-erasable. This attribute ensures that data cannot be accidentally (or intentionally) deleted or modified. Whilst regarded as a disadvantage in some applications, this property is ideal for companies intending to distribute or publish data. It is also ideal for preserving and archiving raw data that should not be altered or eradicated.

More recently, recordable CD-ROM discs have appeared (CD-R). Using a special CD-ROM drive and publishing software, an ISO 9660 image can be prepared and 'written' using a high-powered laser to a blank CD-R disc. The resulting disc is, to all intents and purposes, a CD-ROM conforming in every way to those manufactured. Unlike full manufacture, however, single or low numbers of discs can be produced very economically. CD-Rs offer an enormous cost saving for the low-volume production of CD-ROMs. The method of production does not confer the same durability as those prepared in the traditional way. The estimated life span of a CD-R disc is in the order of 10–15 years or more. Nevertheless, many of the advantages are retained, such as being non-magnetic, non-erasable, not subject to minor scratches, etc. Of course, the ISO 9660 standard assures readability in any drive.

## Reliability

The early definition of the CD-ROM also needed to address the problem of ensuring data quality. Even the most discerning listener is oblivious to minor errors in replaying audio signals from a compact disc. Computers, on the other hand, are not so forgiving; a single corrupted byte could render an entire file useless. Any type of media can become subject to errors when data are read back. Most media have an in-built mechanism whereby small errors can be corrected automatically when reading the data. The ISO 9660 format, therefore, also defined the method of error detection and error correction to validate the data read from the CD-ROM. The potentially large amount of storage space available enabled the designers to be extravagant with their error detection methods. For every 2048 bytes of data stored, 304 bytes are added for error correction. Claims for the precise error rate range between $10^{-18}$ and $10^{-25}$. However, as a minimum it is calculated that 1 byte in 1 trillion megabytes is achievable. This is equivalent to 1 byte in 2 billion CD-ROMs (a stack approximately 2500 km high). This compares with error detection in data transmission using modems of only 1 byte in 10 000 bytes or on standard magnetic tapes of 1 byte in 100 000 bytes.

**Table 1.** *Relative distribution costs of different media (Helgerson 1990)*

| | |
|---|---|
| Paper | US$7/Mb |
| On-line data access | US$200/Mb |
| Floppy disc | US$2/Mb |
| Hard disc | US$20/Mb |
| CD-ROM | US$0.005/Mb |

## Usability

From a commercial aspect, the low weight and size and rapidly reducing mastering costs for high-volume production make CD-ROM the ideal method of distributing large quantities of data, text and information. The common standard also enables a wide range of users to assimilate the information and the robustness and durability of the media ensures a very low level of damage in transit. Replication costs can be as low as that of floppy discs but with 500 times more data. Relative distribution costs are compared in Table 1.

Although perhaps less significant today compared with removable optical discs and so-called 'flopticals', the early CD-ROMs offered unprecedented data storage on a very compact and convenient medium.

Data are stored on the CD-ROM on a track beginning in the centre as a series of 'pits', almost 2 billion on each CD-ROM. The spaces between the pits (or 'lands') are 'read' using a laser beam and photo-sensor measuring the amount of light reflected light. The audio parentage of the CD-ROM is still apparent today with capacity measured in minutes rather than bytes. Up to 79 min of playing time may be recorded on a CD-ROM; this refers to the time taken to follow the spiral track from beginning to end. Each 0.013 s can record 2048 bytes of data, giving a maximum capacity of 746 megabytes. Few CD-ROMs contain more than 600 megabytes, however, since many CD-ROM drives are unable to read data from the extreme outer edge of the CD-ROM. Even so, the CD-ROM capacity is capable of storing over 300 000 pages of text or the equivalent of 130 nine-track 1600 BPI tapes.

## Applications of CD-ROM

CD-ROMs are already in evidence within the field of geology and specifically within the petroleum industry. Several datasets have already been published in the United States including the results from the Deep Sea Drilling Project (DSDP) and the Ocean Drilling Project (ODP). These comprise vast quantities of data that have been collated and are easily accessed from CD-ROMs. Access to these datasets is provided using simple access programs running under MS-DOS® that present the user with maps of the world showing the location of all DSDP borehole locations (Fig. 1). A 'box' outline is moved around the screen to select those boreholes of interest. A selection of available datasets is then provided (Fig. 2), from which further specific selections can be made and viewed on screen or output as a report or file. Other geophysical datasets ranging from gravity and magnetic anomaly surveys to seismic surveys and earthquake data are also available (see Appendix 2). Most of these CD-ROMs have been published by government departments or agencies where the data are public domain and publication is on a non-profit-making basis.

Service companies have also sought to employ the benefits of CD-ROM for the distribution of large amounts of data on a commercial basis. Probably the earliest of these was the Mundo-Cart CD-ROM produced by Petroconsultants in 1987. This CD-ROM contained detailed coastlines, rivers, major towns and other topographic features from 275 maps (1 : 1 000 000) worldwide and was distributed with retrieval and display software for MS-DOSã based machines. The data were regularly updated and have been utilized both inside and outside the petroleum industry as a source of global digital base maps.

Another early commercial example was produced by Erico PI Limited as a means of distributing digitized wireline log data for wells from their North Sea Digital Log Database. The first CD-ROM contained data for 500 released UK continental shelf wells; however, later editions comprised data for almost 1000 wells on a single CD-ROM. Full log suites were supplied sampled at standard six-inch intervals. Access software was included as a Microsoft Windows™ Version 2.3 application enabling users to select wells and download data from the

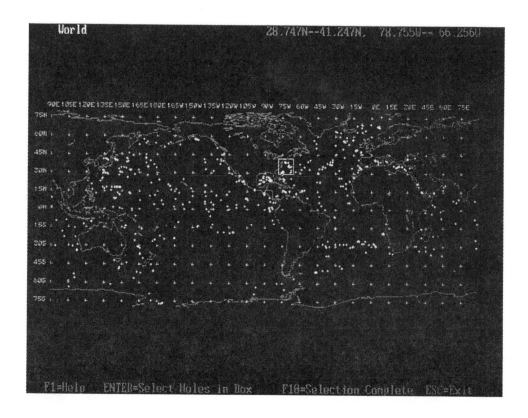

**Fig. 1.** Screen display of map showing locations of DSDP boreholes.

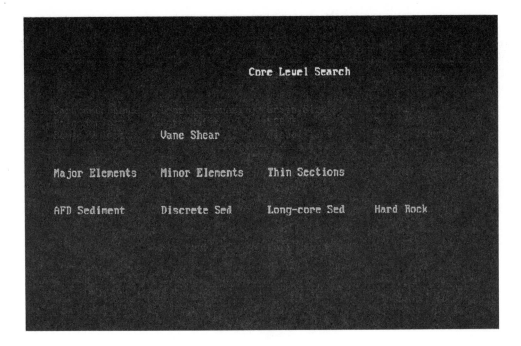

**Fig. 2.** Example of datasets available on the DSDP CD-ROM database.

CD-ROM to conventional disc. The location of wells in the database could be selectively displayed on a map of the North Sea. Using a mouse, the operator was able to point at wells and view a catalogue of curves and intervals available (Fig. 3). Individual curves could then be selected, again using the mouse, for decoding and reformatting. In order to protect Erico PI and the purchaser from unauthorized access to the data, a username and a password were required before data could be downloaded from the CD-ROM. The executable programs contained a unique embedded code that was checked against the password, adding further security.

In the previous two commercial examples the high value of the data more than compensated for the high cost involved in producing a CD-ROM. As the cost of CD-ROM manufacture has decreased, so the commercial viability of publishing data in this fashion has increased.

In 1989 the AAPG (American Association of Petroleum Geologists), in association with Masera Corporation, embarked upon a project to publish AAPG datasets and publications in digital form. Masera–AAPG Data Systems was created and began scanning and processing text,

figures and tables from various AAPG publications. Several projects were begun intended to produce digital datasets on floppy discs or tapes. The most ambitious of these projects was undoubtedly the creation of ultimately three CD-ROM volumes for Microsoft Windows 3.0[TM] and Macintosh[TM] comprising the text, figures and tables for the International Developments Issues 1950 to 1990. Completed in 1992 and delivered with full-text retrieval software, this marked the first of hopefully many such publications including ultimately the bulletin itself on CD-ROM.

Masera–AAPG Data Systems were not alone in pursuing this technique to publish geological papers on CD-ROM. Integrated Digital Services Limited (IDS) in London attempted to create an Earth Science Digital Library[TM] comprising geological papers in digital form from around the world. The first of these, titled the 'Geology of the Arabian Peninsula', was based upon papers published by the USGS (United States Geological Survey). It was delivered with a commercially available Microsoft Windows[TM] Version 3.0 based full-text retrieval software called MediaBase[TM]. In addition to the complete text, the figures, plates and tables were also

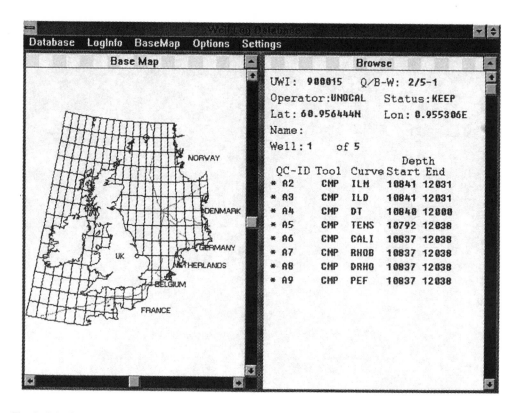

**Fig. 3.** Selection screen from early well log database on CD-ROM produced by Erico PI Limited (1986).

captured as bitmap or spreadsheet files. Microsoft Windows™ based application software, such as Microsoft Excel™ and Paintbrush™ can be accessed from within MediaBase™ and the relevant file automatically loaded from the CD-ROM disc.

An initial selection screen enables users to search the database using titles or any word contained in the text. Searches may be limited to an individual paper or the entire database (Fig. 4).

More complex searches can also be performed using words, combinations of words, proximity of words, full Boolean logic, for example 'sandstone and shale but not limestone', and even partial word searches. Figure 5 shows an example where the user has selected any reference to the words Upper Jurassic within 100 words of any word beginning 'sand', that is sandstone, sandy, etc.

A list of section titles which all conform to the search criteria is presented. Selection of one of these will display the full text on screen with the searched text highlighted in colour (Fig. 6).

'Hot keys' are also embedded in the text where references to other sections or figure captions are present. Using Microsoft Windows Dynamic Linked Library (DLL), calls to other applications can be made. Using icons on the screen, users can retrieve and view in an appropriate application, such as Paintbrush™, the scanned images of diagrams, maps, sections or photographs. Figure 7 shows a section of map automatically viewed from within the application by pointing to the reference in the text.

Numerical tables and statistics are stored as spreadsheet tables. Using the icons and the DLL functions of Microsoft Windows™, the spreadsheets can be retrieved and loaded into an application such as Excel™. Charts and calculations can then be performed (Fig. 8).

Text retrieval software is also employed to provide access to reference databases such as the GEOREF® CD-ROM from the American Geological Institute (AGI). Simple MS-DOS® based programs are used to access the data. In the example described a subset of the GEOREF®

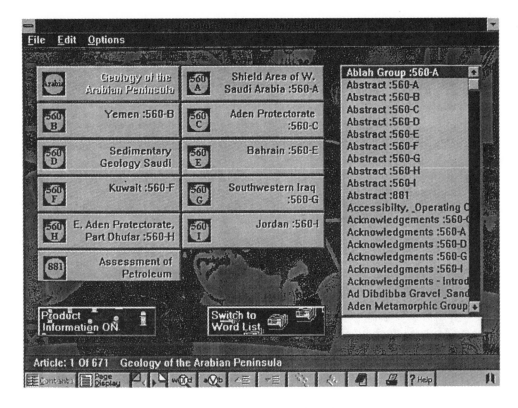

**Fig. 4.** Initial search screen for Geology of the Arabian Peninsula (IDS 1991).

database containing only those entries relating to USGS personnel has been used. Entering a word or combination of words (Fig. 9) results in a list of titles (Fig. 10) and ultimately the full reference and abstract (Fig. 11) being retrieved from the disc.

The reduced cost of CD-ROM production together with its long shelf-life and international standards has also encouraged some companies to view CD-ROM in quite a different light. Instead of using CD-ROM as an inexpensive distribution method, its durable nature is employed as a long-term storage medium for archiving digital data for the future. The ever-reducing mastering costs and more recently the introduction of CD-Recordable (CD-R) drives has made limited runs of CD-ROMs or even individual copies cost-effective. In 1991, British Gas plc became the first major oil company to initiate the systematic archiving of all its digital electric log data tapes to CD-ROM. Other oil and gas companies have followed this example with several in the UK adopting this strategy. These data are held as copies of the original

tapes and commonly utilize the Tape Image Format as defined by Western-Atlas Inc. This format, also referred to as Atlas TIF, PCTIF or LISTIF, preserves the structure and contents of the original tapes on the CD-ROM. Further impetus for the use of CD-ROMs for archiving has been provided by the UK Department of Trade and Industry (DTI). A recent agreement was reached between the DTI and Simon Petroleum Technology Limited to collate and jointly publish digital log data from released UK continental shelf wells. This agreement specified that copies of the original data should be delivered to the DTI on CD-ROM and that the compilation product should also be available on CD-ROM. A similar agreement with Integrated Digital Services Limited includes the compilation and archiving of other wireline data such as Dipmeter where high sample rates make digitization impossible.

It is not only digital log data that are being archived onto CD-ROM. The UK DTI also jointly published with Integrated Digital Services Limited the first ever government release of

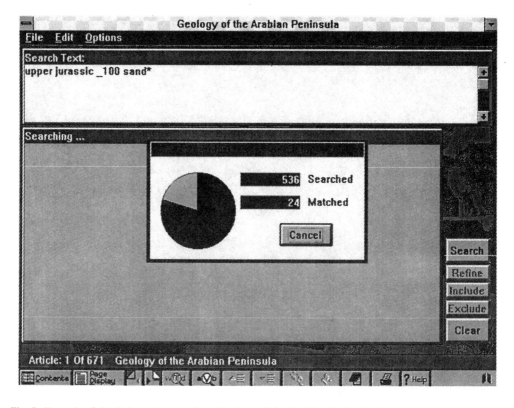

**Fig. 5.** Example of detailed word search from Geology of the Arabian Peninsula (IDS 1991).

seismic data in the UK. Again this was published on CD-ROM and comprised seismic line and navigation data from a survey shot in the offshore southern North Sea. Subsequently Lynx Information Systems Limited also utilized CD-ROM as a means of delivering released seismic data from offshore Netherlands and the western approaches of the UK offshore.

In the summer of 1992 the concept of using CD-ROM as a data storage and distribution medium was extended still further by a major UK oil company in association with Integrated Digital Services Limited. This application included scanned images of core photographs, text reports, floppy disc files, tapes, indeed all relevant data for a single well archived onto an individual CD-ROM disc. The end product provided a convenient means of assembling and distributing to partners all the data for a well. Furthermore, the permanent archiving of all data on a single disc reduces storage costs and prevents data loss or subsequent miss-filing. Since this time, other datasets have emerged on

CD-ROM disc. Further data released and jointly published by the UK DTI and Simon Petroleum Technology Limited comprised core analysis results including digital data and scanned images of the reports. Commitment to CD-ROM media has also been demonstrated by British Petroleum plc who have contracted Simon Petroleum Technology Limited to validate and transcribe their entire wireline log library from conventional magnetic tapes to CD-ROM discs. These discs are then held in a juke-box accessed remotely by the users.

In 1993 another step was taken by the DTI and Sovereign Oil and Gas Limited when it was agreed to publish all material from a relinquished licence block in the northern North Sea. This dataset includes logs, reports, seismic, navigation data, licence application documents, core photographs, special analysis reports and any other related data. Again a prerequisite to the publication was that all material would be distributed on CD-ROM disc as either digital data or scanned images.

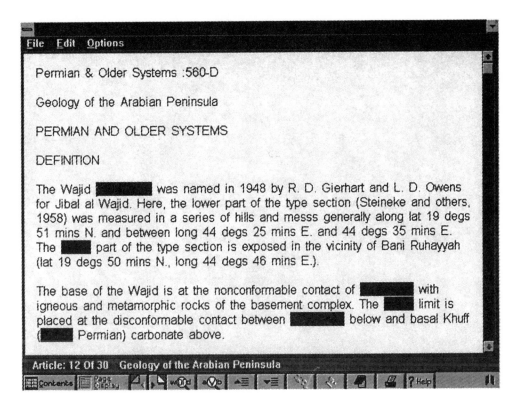

File  Edit  Options

Permian & Older Systems :560-D

Geology of the Arabian Peninsula

PERMIAN AND OLDER SYSTEMS

DEFINITION

The Wajid ▮▮▮▮▮▮ was named in 1948 by R. D. Gierhart and L. D. Owens
for Jibal al Wajid. Here, the lower part of the type section (Steineke and others,
1958) was measured in a series of hills and messs generally along lat 19 degs
51 mins N. and between long 44 degs 25 mins E. and 44 degs 35 mins E.
The ▮▮▮▮ part of the type section is exposed in the vicinity of Bani Ruhayyah
(lat 19 degs 50 mins N., long 44 degs 46 mins E.).

The base of the Wajid is at the nonconformable contact of ▮▮▮▮▮ with
igneous and metamorphic rocks of the basement complex. The ▮▮▮ limit is
placed at the disconformable contact between ▮▮▮▮▮ below and basal Khuff
(▮▮▮ Permian) carbonate above.

Article: 12 Of 30   Geology of the Arabian Peninsula

**Fig. 6.** Results of detailed query from Geology of the Arabian Peninsula (IDS 1991).

## Summary and conclusions

CD-ROMs have emerged as a by-product of the audio industry, where their success has been overwhelming. In the early days of CD-ROM they were described as a solution looking for a problem; to many sceptics this is still the case. Through this paper it is hoped that sceptics will be convinced that CD-ROM does indeed have a place in the petroleum industry. Furthermore, many companies and government agencies are already realizing some of this medium's enormous potential. As a distribution medium, CD-ROMs can deliver large quantities of data cheaply and reliably. For archiving purposes, CD-ROMs can store impressive volumes of data on an extremely durable medium. Furthermore, through the ISO 9660 standard, CD-ROMs are accessible from many different computer platforms, thereby reducing the problem of redundant technology and enabling users to upgrade their equipment without needing to reformat

their archive. For text-retrieval databases the capacity and convenience of CD-ROMs mean that they will increasingly be used for digital books and publications.

It is widely believed that CD-ROMs and readers will become as commonplace and essential to the petroleum company as tapes and tape-drives were in the 1970s and 1980s. Many manufacturers of hardware and software, including Sun, Digital, Hewlett-Packard, Silicon Graphics, Intergraph, Macintosh, Microsoft, Lotus and others, are already committed to supplying software upgrades and documentation on CD-ROM in preference to other media and manuals. This has resulted in CD-ROM drives becoming a standard peripheral on most computer configurations and is generating the demand for more data and titles on CD-ROM.

The author visualizes an industry where publications, both professional society journals and commercial reports, will be delivered as text databases on CD-ROM. Digital data from all

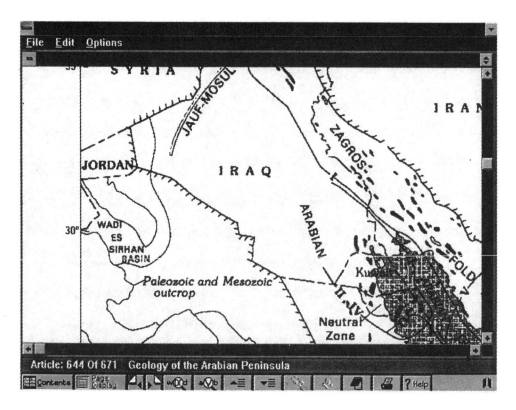

**Fig 7.** Example of scanned image of map viewed from within the Geology of the Arabian Peninsula database (IDS 1991).

disciplines will be archived, exchanged and delivered on CD-ROMs as standard. Individual training and tuition will be provided using multi-media and interactive CD-ROMs, enabling staff and students to learn at their own pace and to simulate the effects of their actions.

Appendix 2 contains a selected list of currently published CD-ROMs pertinent to the petroleum industry and the name and address of the publishing company from where further information can be obtained.

I should like to thank the following companies and people for providing information and assistance in the preparation of this paper: W. T. C. Sowerbutts of the University of Manchester, C. Moore of National Oceanic and Atmospheric Administration, Attica Cybernetics Ltd and Nimbus Information Systems plc.

## Reference

HELGERSON, L. W., (ed.) 1990. *Disc Magazine*, Premier Edition, 1990.

## Appendix 1

### History of the CD-ROM

Optical discs date back to the 1960s when laser technology was first demonstrated; by the 1970s lasers were being used to read data from pits or depressions etched into discs. In the 1980s WORM (Write-Once Read-Many) discs were developed primarily as archiving or back-up systems. These used a laser beam to 'burn' the pits into a substrate for later reading using a lower-powered laser. Later developments produced the REO (Removable Erasable Optical) or MO (Magneto-Optical) disc. These differed from WORM discs by using a combination of optical and magnetic technology, enabling data to be repeatedly written and erased from the disc. More recent developments utilize the changing reflectivity of the media substrate when exposed to high-powered lasers to create erasable optical discs with no magnetic component at all. The REO technology has ultimately

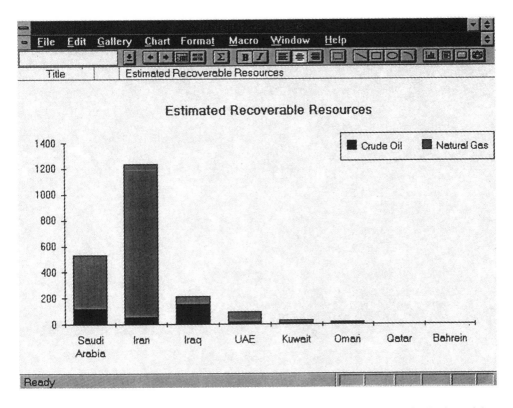

**Fig. 8.** Histogram chart created from statistics contained in a spreadsheet accessed from the Geology of the Arabian Peninsula database (IDS 1991).

spawned a whole range of products aimed at providing high-capacity removable discs. These discs range in capacity from 20 megabytes to hundreds of megabytes or even gigabytes.

CD-ROM is an acronym and stands for Compact Disc Read-Only Memory. CD-ROMs have their roots in the audio industry where, in the early 1980s, laser technology in the form of compact discs (CDs) was employed to replace traditional records and cassette tapes. In 1984 Phillips promoted the concept of using CDs to store and distribute digital data for use with computers. In May 1985 a group of interested parties including software, hardware and media producers met to define the CD-ROM and how data should be stored on it. The result became know as the High Sierra Format, named after the High Sierra Hotel where the meeting was held. In 1988 the High Sierra Format finally became accepted as the standard and, with some slight modifications, was ratified by the International Standards Organisation in 1989 and published as the ISO 9660 format. Several standard publications now exist which define the method of storing audio (Red Book), digital data (Yellow Book), multi-media (Green Book) and Recordable CD-ROM, CD-R (Orange Book) on compact discs. A new addition to these is the so-called White Book, which defines how full motion pictures and video can be compressed and stored on CD-ROM.

## Description of a CD-ROM

The standard CD-ROM is 120 mm in diameter (4.72 inches), although 3 inch CD-ROMs are becoming available for certain applications, 1.2 mm thick, single-sided and contains a spiral track approximately 3 miles long, 0.6 μm wide, 0.12 μm deep and 1.6 μm apart. This spacing corresponds to a track density of 16 000 tracks per inch (TPI), which is much higher than that of floppy discs (96 TPI) or Winchester discs with only several hundred tracks per inch. The CD-ROM revolves at 530 rpm reducing to 230 rpm

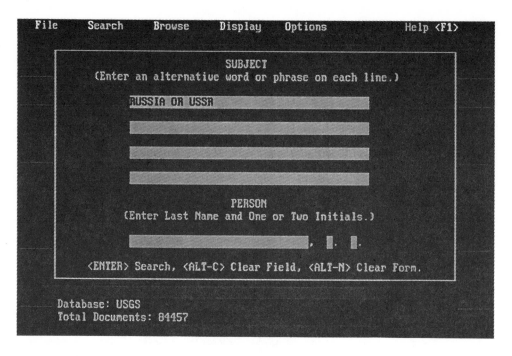

**Fig. 9.** Search screen from the USGS version of GEOREF® on CD-ROM.

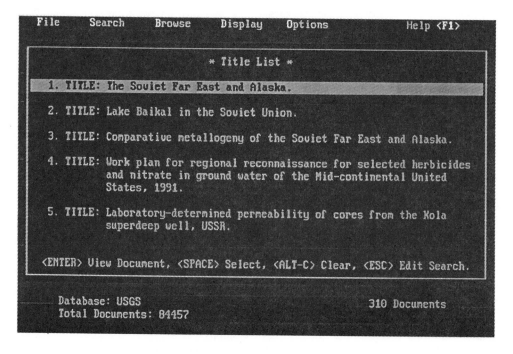

**Fig. 10.** List of titles retrieved from USGS version of GEOREF® on CD-ROM.

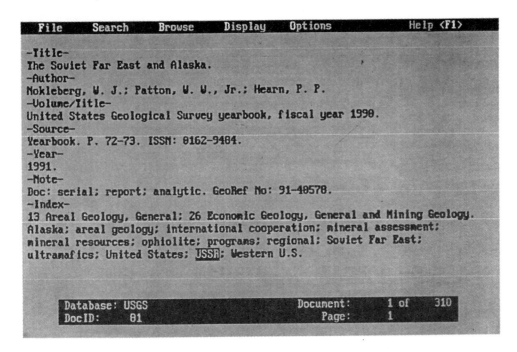

| File | Search | Browse | Display | Options | | Help <F1> |
|------|--------|--------|---------|---------|--|-----------|

-Title-
The Soviet Far East and Alaska.
-Author-
Nokleberg, W. J.; Patton, W. W., Jr.; Hearn, P. P.
-Volume/Title-
United States Geological Survey yearbook, fiscal year 1990.
-Source-
Yearbook. P. 72-73. ISSN: 0162-9484.
-Year-
1991.
-Note-
Doc: serial; report; analytic. GeoRef No: 91-40578.
-Index-
13 Areal Geology, General; 26 Economic Geology, General and Mining Geology.
Alaska; areal geology; international cooperation; mineral assessment;
mineral resources; ophiolite; programs; regional; Soviet Far East;
ultramafics; United States; USSR; Western U.S.

| Database: USGS | Document: | 1 of | 310 |
|----------------|-----------|------|-----|
| DocID: 81 | Page: | 1 | |

**Fig. 11.** References and abstracts retrieved from USGS version of GEOREF® on CD-ROM.

as the reading head moves towards the outside, maintaining a constant linear velocity and therefore data transfer rate. This differs from other optical discs, such as WORM and REOs, which are formatted into sectors and tracks similar to conventional magnetic discs and maintain a constant angular velocity of the reading head.

The ISO 9660 format determined the physical dimensions and characteristics of the CD-ROM, making all CD-ROM drives compatible. Furthermore, and equally important, the format established a method of encoding data that was independent of the hardware and operating system used to create it. This format is non-operating system specific; the software drivers installed on a PC, for example, translate the ISO 9660 format into familiar directories and file names. The same disc attached to a Macintosh™, however, appears familiar to Macintosh users as folders and documents. Provided that the file-naming convention used is based on the lowest common denominator, i.e. MS-DOS®, and the data files are generic, i.e. stored as ASCII or other standard file formats, then the disc is useable on any hardware and operating system compatible with the ISO 9660 format.

Data are stored on the CD-ROM on the track beginning in the centre as a series of 'pits', almost 2 billion on each CD-ROM, and 'lands', the spaces between the pits. These are 'read' using a laser beam and photo-sensor measuring the amount of light reflected light. The changes in light intensity as the beam crosses from a pit to a land, or vice versa, are interpreted as binary '1's and the length of the pit or land represents a series of binary '0's. Up to 79 min of playing time may be recorded on a CD-ROM; this refers to the time taken to follow the spiral track from beginning to end. Each 0.013 s can record 2048 bytes of data, giving a maximum capacity of 746 megabytes.

The ISO 9660 format also defines the method of error detection and error correction to validate the data read from the CD-ROM. For each 2048 bytes of data stored, 304 bytes are added for error correction. Precise error rates range between $10^{-18}$ and $10^{-25}$; however, as a minimum it is calculated that 1 byte in 1 trillion megabytes is achievable, equivalent to 1 byte in 2 billion CD-ROMs. This compares with error detection in data transmission using modems of only 1 byte in 10 000 bytes or magnetic tapes of 1 byte in 100 000 bytes.

## Variations of CD-ROM

CD-ROM has now developed into an entire range of derived products and variations. Most of these have been developed to deliver multimedia datasets including data, sound, graphics and even motion video. A bewildering array of acronyms and buzz-words has evolved to confuse the layman and threatens to undermine the vary cornerstone of CD-ROM technology, i.e. the unparalleled standardization and uniformity of format. Some of these variations are defined below.

CD-A — Audio Compact Disc, 74 min of digital stereo; this represents the standard compact disc purchased for domestic use.

CD-3 — 3-inch Audio Compact Disc, 20 min of digital stereo. This has not been successful in the audio industry, however, the introduction of the Sony Data Disc man which uses 3-inch discs may well provide renewed interest.

CD+G — Audio Compact Disc with graphics for standard television.

CD+EG — Audio Compact Disc with Enhanced (256 colours) Graphics for television.
(CD+G and CD+EG both require special compact disc players to play back through the television and for this reason have not been successful.)

CD-I — Compact Disc Interactive, intended to combine audio with graphics, possibly even motion video, through standard television. Unlike CD+G, CD-I enables users to interact and select options from menus, query databases, etc. This also requires a modified compact-disc player capable of reading CD-I discs.

CDTV — Commodore Digital Total Vision, this is a direct competitor of CD-I, launched by Commodore in 1991. Current titles include encyclopaedias, games, etc., and it is designed for domestic and educational use.

Photo-CD — Compact Disc containing photographs and created during development to produce digital images on disc. This has been developed by Kodak and is designed to hold up to 100 photographs at various resolutions for display on domestic television and computers. This also requires a CD-I compatible player to read the discs.

CD-ROM XA — CD-ROM extended architecture allows multi-media audio, graphics and still photographs to be combined with data and played back through a computer.

CD-R — Compact Disc Recordable or CD-WORM discs, used initially to produce test CD-ROMs prior to full manufacture. The introduction of less-expensive drives has enabled CD-ROMs to be produced in very small numbers, even individual CD-ROMs for archiving and data storage purposes.

Multi-Session — A Multiple session CD-Rs are currently emerging which are capable of being updated by allowing additional image files to be recorded at the end of an existing CD-ROM track, space permitting. Players capable of reading these additional sessions were only launched recently; many of the currently installed CD-ROM drives are unlikely to be able to access these other sessions.

CD-E — Future development to produce erasable CD-ROM (CD-MO). The technology has not yet been perfected; however, as with most things, it is probably just a question of time.

CD-EP — Extended-play CD-ROMs are already possible with a capacity of over 2 gigabytes. However, drives capable of reading these CD-ROMs are not yet in production. The additional capacity is gained by reducing the gap between the tracks, thereby increasing the number of tracks per inch. This requires a narrower focused laser beam which is not possible with the current gallium arsenide lasers in existing CD-ROM drives.

DVI        Digital Video Interactive is not a type of CD-ROM but a patented processing technique for digital video. This method requires a special video card developed by Intel to compress and decompress digital video images in real time to enable motion video to be played back.

## Manufacture of CD-ROM

In order to appreciate some of the advantages and disadvantages of CD-ROM it is helpful to understand the processes required to prepare and manufacture them. CD-ROMs cannot be 'written to' like other discs and tapes (except for CD-R discs). In principle, they are made in a similar way to audio records, requiring expensive capital equipment designed to produce hundreds, thousands or even millions of copies.

There are essentially three stages required: data preparation, pre-mastering or formatting and mastering. Before the development of the CD-R recorders, all CD-ROMs were mastered at specialist factories. These facilities were created to service the audio industry, producing millions of CD-ROMs for domestic use. Costs to produce short-runs of CD-ROMs for publication of books or data was relatively expensive (several thousand dollars). It was essential, therefore, that every care was taken in preparing the data since mistakes could not be rectified after mastering. Furthermore, it was obviously an advantage to defer mastering until sufficient data existed to utilize all the space on the CD-ROM. This resulted in long product development times and high data-preparation costs.

Once the design and preparation are complete the entire contents of the CD-ROM including data, software and layout is converted into a single CD-ROM image. This process, known as pre-mastering or formatting, is often performed by the mastering company, however, increasingly CD-ROM authors are acquiring the necessary software to perform this stage themselves. Once formatted, a single file is created which can be tested as a simulated CD-ROM for evaluation. Normally this image will conform to the ISO 9960 standard; however, CD-ROMs can also be prepared specifically for UNIX[TM], VMS[TM] and other operating systems. It should be stressed, however, that only the ISO 9960 format is non-specific to the host system.

When the image is complete and tested, a final CD-ROM can be made. Using the recent CD-R drives the image can be burnt into a blank master to produce a single CD-ROM. This CD-ROM behaves exactly the same as conventionally prepared CD-ROMs. However, CD-Rs are generally more vulnerable to damage and corruption, with an estimated shelf-life of 12–15 years compared to a manufactured CD-ROM with an estimated shelf-life of over 35 years. Alternatively, and more commonly, the CD-ROM image is used to manufacture conventional CD-ROMs using a lengthy and costly process.

First a master disc, usually glass, is coated with a film of photo-resistant chemical. The ISO 9660 image file is transferred to the master using a blue light laser. The master is then etched in developer, leaving pits in the photo-resistant layer where it has been exposed to the laser. This master disc is coated with a thin film of silver and tested for faults. The silver-coated master is electroplated with nickel and removed from the glass giving a mirror image copy or stamper. The stamper is attached to a mould into which polycarbonate plastic is injected to form the final CD-ROM. The CD-ROM is coated with aluminium to provide a reflective surface, and sealed and cured in ultraviolet light to prevent corrosion and scratching. Finally, the label or artwork is printed onto the top surface of the CD-ROM. The entire process can take as little as a day, although normal processing turnaround is 5–10 days. Manufacture is essentially a production line and a minimum order of 10–50 CD-ROMs is often required.

**Appendix 2**

*Selected list of available titles and publications available on CD-ROM*

| Title | Publisher | Description |
|---|---|---|
| *AAPG Enhanced Bulletin 1991* | Masera-AAPG Data Systems | Full text, figures and tables for all 1991 AAPG bulletins |
| *AAPG International Developments* | Masera-AAPG Data Systems | Full text, figures and tables for International Developments issues from 1950 to 1990 |
| *AAPG Oil and Gas Fields— A Digital File of Case Histories* | Masera-AAPG Data Systems | Full text, figures and tables for 78 oil and gas fields, based on Treatise of Petroleum Geology Atlas of Oil and Gas Fields |
| *Aerial Photography Summary Record System* | USGS | Aerial photographs taken by over 500 federal, state and private industry |
| *Deep Sea Drilling Project* | NOAA | Results from first 99 legs of the DSDP project including wireline logs, core descriptions and analyses, geochemistry, palaeontology, bio-, chrono- and lithostratigraphy |
| *Earth Sciences* | OCLC | Includes Earth Science Data Dictionary, GEOINDEX and USGS library Catalogue database |
| *Earthquake Data* | USGS | Data for earthquakes greater than magnitude 5.5 beginning 1980 |
| *Energy Library* | OCLC | Bibliographic reference for oil, coal, nuclear, solar and other related energy sources |
| *Experimental Calibrated GVIs* | NGDC | Normalised Difference Vegetation Index (NDVI) based on Advanced Very High Resolution Radiometer |
| *GEODAS* | NGDC | Geophysical Data System, worldwide marine geophysical data. Bathymetry, magnetics, gravity, and seismic shot-point navigation |
| *GEOINDEX* | OCLC | Guide to published maps of the US and its territories |
| *Geology of the Arabian Plate— Volume 1 Earth Science Digital Library* | IDS Ltd | Full text and figures from ten USGS publications covering the geology and petroleum potential of the region |
| *Geophysics of North America* | NGDC | Topography, magnetics, gravity, satellite imagery, seismicity, crustal stress, thermal aspects and geopolitical boundaries |

**Appendix 2** *(continued)*

| Title | Publisher | Description |
|---|---|---|
| *GEOREF* | AGI | Bibliographic database of earth science references and abstracts |
| *Global Ecosystems Data* | NGDC | Vegetation, climate, topography, soils and other data on global ecosystems |
| *GLOTIA Sidescan Sonar Data* | USGS, NOAA, NASA | Images of the Gulf of Mexico sea bed |
| *Marine Minerals* | NOAA | Results of a nine-year effort to collate bibliography and geochemical database of offshore hard mineral resources |
| *Middle East—Regional Geology and Petroleum resources* | Masera-AAPG Data Systems | Full text and figures of book of the same title written by Z. R. Beydoun (Scientific Press Ltd) |
| *Minnesota Aeromagnetics* | NGDC | Aeromagnetics data for Minnesota |
| *MUNDOCART/CD* | Petroconsultants | Contains mapping data points for whole world, covering more than 500 000 features |
| *NEIC Database* | National Earthquake Information Centre | Seismic records of earthquake information worldwide |
| *NGDC-01* | NGDC | Selected geomagnetic and other solar–terrestrial physics data of NOAA and NASA |
| *North Sea Project* | BODC | Marine data including trace metals, inorganic and organic atmospheric chemistry, salinity, phytoplankton and other data from the North Sea |
| *Ocean Drilling Project* | NOAA | Results from Legs 101–135 of the ODP project including wireline logs, core descriptions and analyses, geochemistry, paleontology, bio-, chrono- and lithostratigraphy |
| *Publications of the USGS* | AGI | USGS subset of the GEOREF bibliographic database, 1845 to present |
| *STP* | NGDC | Flares in hydrogen-alpha and regions of solar activity from ground-based observations |

* Publishers
(AGI) American Geological Institute, GeoRef Information System,4220 King Street, Alexandria, Virginia, 22302–1507, USA
(BODC) British Oceanographic Data Centre, Bidston Observatory, Birkenhead, Merseyside L43 7RA, UK
(IDS) Integrated Digital Services Limited (see Simon Petroleum Technology) Masera-AAPG Data Systems, 1743 East 71st Street,Tulsa, Oklahoma 74136, USA

(NASA) National Aeronautical and Space Administration, California Institute of Technology, Division of Earth and Space Sciences, Pasadena, California 91109, USA

(NEIC) National Earthquake Information Centre, Mailstop 967, P.O. Box 25046, Denver Federal Centre, Denver, CO 80225, USA

(NGOC) National Geophysical data Center, 325 Broadway, E/GC1, Dept 915, Boulder, Colorado 80303-3328, USA

(NOAA) National Oceanic and Atmospheric Administration 325 Broadway, E/GC4, Dept 915, Boulder, Colorado 80303-3328, USA

(OCLC) On-line Computer Library Center, 6565 Frantz Rd, Dublin, Ohio 43017-0703, USA

(CES) Petroconsultants Limited,Burleigh House, 13 Newmarket Rd, Cambridge CB5 8EG, UK

(SPT) Simon Petroleum Technology Limited (incorporating Integrated Digital Services Limited), Tyn-y-Coed, Llanrhos, Llandudno, Gwynedd LL30 1SA, UK

(USGS) United States Geological Survey, National Center, MS 511, Reston, Virginia 22092–9998, USA

# Index